MicroRNA in Cancer

Suresh Alahari
Editor

MicroRNA in Cancer

Springer

Editor
Suresh Alahari
Department of Biochemistry and Molecular Biology
Stanley S. Scott Cancer Center
LSU Health Science center
New Orleans, Louisiana
USA

ISBN 978-94-017-8493-1 ISBN 978-94-007-4655-8 (eBook)
DOI 10.1007/978-94-007-4655-8
Springer Dordrecht Heidelberg London New York

Softcover reprint of the hardcover 1st edition 2013

Printed on acid-free paper

Springer is part of Springer Science+Business Media (www.springer.com)

Preface

Cancer is a complex and multistep process involving the accumulation of multiple changes that eventually transform normal cells into cancer cells. These changes include structural and expression abnormalities of both coding and non-coding genes. Most cancer-related deaths are not caused by primary tumors but by the spread of cancer cells from the original site to distant sites. In the year 1993, Ambros and colleagues first discovered a gene for lin-4, which did not code for protein, in *C.elegans*, and it was named as microRNAs. Since then several microRNAs have been discovered in various organisms. MicroRNAs have regulatory roles in several biological processes. In cancer, microRNAs function as regulatory molecules acting as oncogenes and tumor suppressors resulting in them having very significant roles in cancer biology. Thus when Springer asked me to work on this book, I accepted the invitation without any second thoughts. Many outstanding investigators have done great amounts of work on microRNA in cancer so we could not cover every study because of space limitations for which we apologize. Our understanding of microRNA's role in cancer is great due to the advent of several genetic engineering approaches through making transgenic and knockout animals for microRNAs. Furthermore, several novel therapeutic modalities for microRNA have reinvigorated many hopes for the cure to cancer. In the last few years microRNA research has grown tremendously, allowing us to get closer to the development of microRNA targeted therapies the usage of microRNAs as diagnostic and prognostic markers. Some microRNAs are detected in the plasma of cancer patients and can serve as diagnostic markers, prognostic markers, therapeutic targets, and causal factors in cancers. The novel microRNA based therapies will likely reduce the incidence of death from cancers. In this book, my goal is to comprehensively review the fundamental knowledge of microRNAs in cancer.

This book is composed of eight chapters that give basic information of the role of microRNAs in cancers. The first chapter describes the general functions of microRNAs and other non-coding RNAs in cancers. Here, authors effectively describe the pivotal role of microRNAs in various malignancies. More importantly, the authors introduce novel non-coding RNAs including MALAT1, HOTAIR and others. The second chapter describes how microRNAs regulate cell proliferation in which authors provide a detailed list of microRNAs that are important in cell proliferation and discuss, in detail, various therapeutic approaches describing the restoration of tumor suppressor microRNA expression and suppression oncogenic microRNAs expression. In the third chapter, the author elucidates the importance of microRNAs in cancer stem cells. He elegantly narrates the cancer stem cell hypothesis, shows links between cancer stem cells and epithelial-mesenchymal transition, and depicts the important role of microRNAs in normal as well as cancer stem cells. The fourth chapter describes how microRNAs regulate viral pathogenesis and cancers including the methods by which viruses regulate microRNA and viral microRNAs regulate host genes. The fifth chapter deals exclusively with oncogenic microRNAs and describes how they function in normal cells and in cancer cells. It also discusses the cell specific microRNAs and shows the importance of microRNAs in resistance to chemotherapy and radiation therapy. The sixth chapter mainly focuses on metastasis specifc microRNAs. The seventh chapter highlights the role of microRNAs in Leukemias. Finally, the eighth chapter describes various novel approaches for making small molecule modifiers of microRNAs that can be used as molecular probes or in therapeutics and the various methods of the delivery of such small molecules. This chapter is a completely new twist from the current thinking concerning microRNAs.

The authors have done a fantastic job in presenting these complex topics in an easy, understandable manner. I am very thankful to the authors who have written these chapters and unselfishly assisted me in my first editing of a book. I would also like to thank the staff at Springer Science located in the Netherlands, especially Ilse Hensen for her assistance in this process. Finally, I would like to dedicate this book to my father, the late Venkaiah Alahari, and my mother, Saraswathi Alahari, who have supported me in every step of my life with whatever little resources they had and without their help I would not be the individual I am today.

Contents

MicroRNAs and Other Non-Coding RNAs: Implications for Cancer Patients

Reinhold Munker and George A. Calin

Abstract The discovery of microRNAs (miRNAs) has shed new light on the role of RNA in gene regulation. MiRNAs are small molecules (size, 19–22 nucleotides) that do not encode proteins but interfere with translation and transcription, thereby regulating gene expression. Multiple miRNAs are dysregulated in human cancer, supporting the hypothesis that miRNAs are involved in the initiation and progression of cancer. Prototypic malignancies in which a role for miRNAs has been demonstrated include chronic lymphocytic leukemia, multiple myeloma, cutaneous T-cell lymphoma and mantle cell lymphoma. More research is necessary, but miRNAs have already improved our understanding of the pathogenesis of cancer. MiRNAs measured in bodily fluids, especially plasma, may be useful as biomarkers for cancer. Beyond miRNAs, several thousand other non-coding (also called ultraconserved) RNAs may be important in the pathogenesis and prognosis of cancer. Some ultraconserved non-coding RNAs interfere with signal transduction by modifying chromatin structures, but most are not yet well characterized. MiRNAs and other non-coding RNAs may be useful for the gene therapy of cancer.

1 Introduction

The literature on microRNAs (miRNAs), and especially miRNAs in cancer, has increased exponentially over the last 10 years. Cancer is a frequent disease: at least one third of the population will develop cancer during their lifetimes. Despite progress in early detection, chemotherapy, immunotherapy, radiation and other treatments, most people with advanced cancer will ultimately die of the cancer. Overall, new treatments for cancer with fewer side effects are urgently needed. The discovery of miRNAs and other non-coding RNAs will lead to new biomarkers for determining the diagnosis, prognosis, and treatment response of cancer and may ultimately lead to new treatments for cancer.

G. A. Calin (✉) · R. Munker
The University of Texas MD Anderson Cancer Center, Houston, TX 77030, USA
e-mail: gcalin@mdanderson.org

R. Munker
Louisiana State University, Shreveport, LA 71115, USA

S. Alahari (ed.), *MicroRNA in Cancer,*
DOI 10.1007/978-94-007-4655-8_1, © Springer Science+Business Media Dordrecht 2013

It is clear that miRNAs are dysregulated in cancer. For many types of cancer, miRNA signatures have been established. Some signatures provide prognostic information. The field of miRNAs in cancer was launched when Calin et al. [1, 4] showed that miR-15 and miR-16 were located in a region (chromosome 13q14) frequently deleted in chronic lymphocytic leukemia (CLL). Consequently, the expression of miR-15 and miR-16 in CLL is decreased. Subsequently, based on 218 samples, Lu et al. [2] showed that cancer can be classified according to miRNA expression. Based on a larger collection of samples and using a customized microarray, Volinia et al. [3] published a miRNA signature of solid tumors. In this chapter, we give an update on the role of miRNAs in cancer exemplified by important disease entities (CLL, multiple myeloma, cutaneous T-cell lymphoma and mantle cell lymphoma) and then look further into other recent developments in the field of non-coding RNA. We recently published a general overview of the topic of miRNAs in cancer [5].

Fundamentally, miRNAs are small molecules (approximate size, 19–22 nucleotides) that do not encode proteins. The major function of miRNAs is to regulate gene expression. It has been estimated that 30 % or more of mammalian genes are regulated by miRNAs. Mechanisms by which this regulation occurs involve degradation of messenger RNA (mRNA), chromatin-based silencing and inhibition of translation. MiRNAs are highly conserved between different species. Currently, more than 600 miRNAs are known or generally accepted. About half of all known miRNAs are located in minimal regions of amplification, at common breakpoints associated with cancer or in close proximity to fragile sites or in minimal regions of loss of heterozygosity [5].

The synthesis of miRNAs begins in the nucleus at the stage of pri-miRNA transcripts. Subsequently, these transcripts are cleaved by an RNase III-type nuclease (Drosha) and form hairpin structures of 60–70 nucleotides (pre-miRNAs). Pre-miRNAs are exported into the cytoplasm by exportin. In the cytoplasm, the enzyme Dicer performs further cleavage, which results in an asymmetric intermediate (MiRNA: MiRNA*). The duplex then makes contact with the RNA-induced silencing complex (RISC), where one strand becomes active and functional (repressing translation and degrading mRNA). The inactive strand (marked by an asterisk or star) is generally not considered of functional importance (although there may be exceptions [6]). For a detailed review about the biogenesis of miRNAs, see Krol et al. [7].

2 MiRNAs in Selected Malignancies

Among the myriad studies and publications about the significance of miRNAs in cancer, we will discuss here four diseases that are relevant to our current research. CLL is the most frequent leukemia in Western countries and has become the paradigmatic disease for the involvement of miRNAs in cancer. Multiple myeloma is the second most frequent hematologic malignancy; it involves bones and bone marrow. Cutaneous T-cell lymphoma and mantle cell lymphoma are rare types of T- and B-cell lymphomas with a wide spectrum of clinical presentations and outcomes. In all these diseases and disorders, miRNAs were shown to be important.

2.1 Chronic Lymphocytic Leukemia

A frequent chromosomal aberration in CLL is the homozygous or heterozygous deletion of the chromosomal region 13q14.3. Patients with this deletion often have an indolent or benign clinical course. In 2002, it was shown by Calin et al. [8] that two genes encoding miRNAs (miR-15a and miR-16-1) are located in this region, providing evidence that miRNAs could be involved in the pathogenesis of human cancer [8]. MiR-15a and miR-16-1 map to a 30 kb region between exons 2 and 5 of the DLEU2 gene (which is deleted in these patients). A common hypothesis is that the loss of both miRNAs is an early event in the pathogenesis of CLL.

In a later study, a unique miRNA signature for CLL was defined [9]. The signature of nine miRNAs (eight whose expression was increased, one whose expression was decreased) correlated with somewhat more aggressive disease. This pattern also corresponded to known biologic risk factors for CLL, such as high expression of 70 kDa zeta-associated protein (ZAP70) and unmutated immunoglobulin heavy chain genes.

The role of miRNAs in the predisposition to or inheritance of cancer is another area of research. In support of such a role, mutations of some miRNA genes were found in 11 of 75 patients with CLL. This discovery points to a genetic disposition for cancer in some patients with CLL.

The New Zealand Black mouse model of CLL supports the role of miRNAs in the pathogenesis of CLL. In this model, a 3' point mutation adjacent to miR-16-1 led to reduced expression of miR-16-1 [10]. In a different mouse model, the deletion of the 13q14 minimal deleted region (encoding the DLEU2/miR-15a/16-1 cluster) caused development of indolent B-cell–autonomous and other clonal lymphoproliferative disorders. This deletion recapitulates the spectrum of CLL-associated phenotypes observed in patients [11]. The loss of miR-15a/16-1 accelerates the proliferation of B lymphocytes both in mice and humans by modulating the expression of genes controlling cell-cycle progression. A mouse model for indolent CLL was recently generated by overexpressing miR-29 in B cells. Such Eμ-miR-29 transgenic mice developed CD5 + B lymphocytosis starting at 2 months of age. By 2 years, the percentage of CD5 + B lymphocytes had increased to 100 %, and about 20 % of the mice died from leukemia [12].

Patients with cancer or leukemia often respond to chemotherapy, but later relapse and become resistant. The topic of resistance to cancer chemotherapy is clinically relevant and may involve miRNAs. The phenotype of *in vivo* fludarabine resistance was described as upregulation of miR-18, miR-122 and miR-21 [13]. The authors studied 723 miRNAs in 17 patients with CLL. RNA was harvested from peripheral blood before and after a 5 day course of fludarabine. Nine patients responded clinically, eight patients were classified as resistant. In responding patients, the activation of p53 responsive genes was detected.

Feedback circuitry linking miRNAs, TP53 and the pathogenesis and outcome of CLL was established by Fabbri et al. [14]. For this study, CLL Research Consortium institutions provided 206 blood samples from untreated patients with B-cell CLL. These samples were evaluated for the occurrence of cytogenetic abnormalities, as well as the expression levels of the miR-15a/16-1 cluster, miR-34b/34c cluster, TP53 and ZAP70. The functional relationship between these genes was studied using *in vitro* experiments examining gain and loss of function and was validated in a separate collection of primary CLL samples. In 13q-deleted samples (as mentioned, associated with a favorable prognosis), the miR-15a/16-1 cluster directly targeted TP53 and its downstream effectors. In leukemic cell lines and primary B-CLL cells, TP53 stimulated the transcription of both miR-15/16–1 and miR-34b/34c clusters, and the miR-34b/34c cluster directly targeted ZAP70 kinase.

The interplay between protein-coding genes and miRNAs, as well as other non-coding RNAs, in CLL was reviewed by Calin and Croce [15].

2.2 *Multiple Myeloma*

The first study involving miRNAs in multiple myeloma showed that interleukin-6 induces miR-21 via Stat3 activation. When miR-21 was increased ectopically, the myeloma cells lost their interleukin-6 dependence [16]. Pichiorri et al. [17] in 2008 were first to establish an miRNA expression profile for multiple myeloma by comparing myeloma cell lines with CD138-selected samples from patients with myeloma, samples from patients with monoclonal gammopathy of unknown significance, and normal plasma cells. In these profiles, miR-21, the miR-106b~25 cluster and miR-181a/b measured in patients' bone marrow myeloma cells were overexpressed compared with expression in normal plasma cells. Two miRNAs, miR-19a/b, which are part of the miR17~92 cluster, were shown to interact with the expression of the SOCS-1 gene. In addition, xenograft studies implicated miR-19a/b and miR-181a/b in the pathogenesis of multiple myeloma [17]. This work was recently extended by demonstrating that miR-192, miR-194 and miR-215 (which are often downregulated in newly diagnosed multiple myeloma) are part of an autoregulatory loop with MDM2 and p53. It was shown that through small-molecule inhibitors of MDM2, these miRNAs can be transcriptionally activated by p53 and then modulate MDM2. In addition, miR-192 and miR-215 target the IGF pathway, preventing the homing of myeloma cells [18]. The correlation between miRNA expression, DNA copy number changes and gene expression was studied by Lionetti et al. [19]. A new histone deacetylase inhibitor (ITF2355) was shown to downregulate miR-19a and miR-19b [20]. In 15 patients with relapsed or refractory myeloma, a decrease of miR-15a and miR-16 and an increase of miR-222, miR-221 and miR-382 were found [21]. In a larger study involving 52 newly diagnosed patients, a global increase in miRNA expression was observed in high-risk disease. High-risk disease was defined by a

70-gene risk score and the proliferation index. Of particular interest is that one of these genes, EIF2C2/AGO2, is considered to be a master regulator of the maturation and function of miRNAs. When EIF2C2/AGO2 was silenced, the viability of multiple myeloma cell lines decreased dramatically [22].

2.3 *Cutaneous T-Cell Lymphoma*

Recently, a miRNA expression profile for Sézary syndrome, the leukemic form of cutaneous T-cell lymphoma, was established [23]. Sézary syndrome generally has a poor prognosis. Most miRNAs expressed in Sézary syndrome were downregulated and distinguished Sézary syndrome both from normal CD4-positive T cells and from B-cell lymphomas. The authors showed that downregulated miR-342 inhibits apoptosis, thereby suggesting a role for this miRNA in the pathogenesis of cutaneous T-cell lymphoma. The work on cutaneous T-cell lymphoma needs to be extended. Of special interest will be whether an *in vivo* resistance profile to histone deacetylase inhibitors can be determined.

2.4 *Mantle Cell Lymphoma*

Mantle cell lymphoma has a well-defined chromosomal marker, t(11; 14), which leads to overexpression of cyclin D1. Zhao et al. [24] performed expression profiling for 30 patients with mantle cell lymphoma and found a decrease in 18 miRNAs and an increase in 21 miRNAs compared with levels in normal B lymphocytes. The authors demonstrated that miR-29 inhibits CDK6 protein and mRNA (which are involved in the pathogenesis of mantle cell lymphoma) by binding directly to the 3'-untranslated region of the mRNA. In addition, they showed that cases with the lowest miR-29 levels in lymphoma cells had the worst prognosis [24]. In a different profiling study of miRNA expression in mantle cell lymphoma (involving eight cell lines and a total of 77 patients), increases in miR-106b, miR-93 and miR-25 were demonstrated (among other changes) [25].

3 Soluble MiRNAs

Plasma or serum tumor markers could enable detection of cancer without invasive procedures. Mitchell and colleagues [26] reported that plasma contains stable miRNAs that are protected from endogenous RNase activity. The use of miRNAs as tumor markers was exemplified by their measurements of miR-141 derived from prostate xenografts. The authors also demonstrated that levels of this miRNA could be used to distinguish patients with prostate cancer from healthy controls. In colorectal cancer, one study showed that levels of two plasma miRNAs were significantly increased in comparison to levels in normal individuals (and decreased after surgery). A possible use of such plasma-based miRNAs might be screening for colorectal carcinoma [27]. Plasma miRNAs were also applied to non–small cell lung cancer [28]. The miRNAs miR-200a/b are often overexpressed in biopsies of pancreatic cancer. Li et al. [29] recently investigated serum miRNAs in pancreatic cancer and found that levels of miR-200a/b in most patients were elevated compared with levels in normal controls. The diagnostic value of miR-200a/b is doubtful, however, because a similar elevation was observed in chronic pancreatitis [29]. In a further study, levels of three miRNAs were increased in whole blood from cancer patients; and one of these miRNAs (miR-195) appeared specific for breast cancer [30]. When 12 different types of bodily fluids (from plasma to colostrum) were investigated, miRNAs were found to be ubiquitous. It was speculated that some miRNAs may transmit signals between cells and tissues [31]. Taken together, the studies on soluble miRNAs have mostly been done with small patient numbers and controls and would need standardization before clinical use could be considered. Nevertheless, if the data are reproducible and valid when all controls are integrated in the protocol, plasma miRNAs may be a simple way of diagnosing cancer.

4 Other Non-coding RNAs

Up to 70 % of the human genome is transcribed, but only 2 % of the genes are translated into proteins. Besides miRNAs, which we have begun to understand, there are many other non-coding RNAs (probably more than 6,000), most of which are not well characterized. Some of these molecules may be important in the regulation of gene expression and, by proxy, also in the pathogenesis and progression of cancer. These molecules may also serve as new biomarkers.

We will discuss here the other non-coding RNAs for which an involvement in cancer has been shown. These include long intergenic non-coding RNAs (lincRNAs), such as MALAT-1, HOTAIR and other transcribed non-coding RNAs. By definition lincRNAs are molecules of more than 200 nucleotides in length. One category of lincRNAS, the ultraconserved non-coding RNAs are identical between mouse, rat and man and therefore deemed important in gene regulation. The mechanisms of gene silencing by lincRNAs may involve epigenetic modifications of chromatin within promoter regions [32]. It was recently reported that a large fraction of genomic ultraconserved regions encode a particular set of non-coding RNAs whose expression (similar to miRNAs) is altered in human leukemias and cancers. Ultraconserved regions are frequently located at fragile sites and genomic regions involved in cancer. These non-coding RNAs may be regulated by miRNAs that are abnormally expressed in human CLL [33]. Similar to miRNAs, the non-coding RNAs transcribed from ultraconserved regions are often hypermethylated in human cancer [34]. According to Huarte and Rinn [35], lincRNAs may provide the "missing link in cancer," implying that these molecules could also function as tumor suppressors and tumor inducers and thereby initiate or promote cancer.

4.1 MALAT-1

MALAT-1 was originally isolated by subtractive hybridization from a pool of metastatic lung adenocarcinomas [36]. The MALAT-1 transcript has 8,000 nucleotides, originates from human chromosome 11q13 and is conserved across several species. MALAT-1 is expressed in several normal tissues, such as the pancreas and lung, and overexpressed in metastatic lung cancer. The overexpression of MALAT-1 in early lung cancer predicts ultimate metastasis and death from metastatic lung cancer. The mouse ortholog of MALAT-1, designated as hepcarcin, was found to be strongly expressed in mouse carcinogen-induced liver cancers, as well as human hepatocellular carcinomas [37]. In osteosarcoma, high expression of MALAT-1 corresponded with poor response to chemotherapy [38]. Recently, it was shown that the silencing of MALAT-1 impaired the motility of lung cancer cells, which may explain the role of MALAT-1 in metastasis. The knockdown of MALAT-1 influenced the expression of numerous genes (including CTHRC1, CCT4, HMMR and ROD1, which on their own also influence cell motility) [39].

4.2 HOTAIR

The lincRNA HOTAIR is in the mammalian HOXC locus and binds to and targets the PRC2 complex on the HOXD locus, which is located on a different chromosome. It was recently shown that HOTAIR is overexpressed a hundred—to a thousand-fold in breast cancer metastases. In primary tumors, HOTAIR expression is a powerful predictor of eventual metastasis and death. The enforced expression of HOTAIR leads to a genome-wide re-targeting of PRC2 to an occupancy pattern resembling embryonic fibroblasts and increased cancer invasiveness and metastasis. These findings suggest that lincRNAs have active roles in modulating the cancer epigenome and may become important targets for the diagnosis and treatment of cancer [40].

4.3 H19

The H19 locus is subject to genomic imprinting and produces a 2.5 kb non-coding, spliced and polyadenylated RNA. It was recently shown that this locus acts as an *in vivo* tumor suppressor in several mouse models of cancer [41].

4.4 XIST

XIST is a non-coding transcript involved in X chromosome silencing. Some recent data point toward an involvement in breast cancer, especially in BCRA-1-related cases [42].

4.5 SnaR Family Members

Members of the snaR family of small non-coding RNAs associate *in vivo* with nuclear factor 90 protein. The major human species (snaR-A) has a restricted tissue distribution (brain, testis and some other tissues) and is upregulated in transformed and immortalized cells. In the HeLa cell line, snaR-A is stably bound to ribosomes [43].

4.6 Other Transcribed Ultraconserved Regions

Transcribed ultraconserved regions were shown to be widely expressed in neuroblastomas, and their expression correlates with important clinicogenetic parameters such as N-MYC amplification [44]. Recently, a novel lincRNA in 8q24 was described. This lincRNA has a size of approximately 13 kb. The authors found several single nucleotide polymorphisms that increased the risk for prostate cancer in Japanese patients; they termed this novel RNA "prostate cancer non-coding RNA1". Knockdown of this novel RNA by siRNA decreased both the viability of prostate cancer cells and the transactivation of the androgen receptor [45]. The HULC gene ("highly upregulated in liver cancer") is a non-coding RNA transcribed from human chromosome 6p24.3. HULC is strongly expressed in hepatocellular carcinomas but also to a lesser extent in normal hepatocytes. It was recently shown that HULC is also expressed in liver metastasis of colon cancer but not in primary colon cancers [46]. The PCGEM1 non-coding gene (previously described as prostate-specific and androgen-regulated) was recently shown to play a role in the *in vivo* progression of prostate cancer [47]. In liver cancer, a transcribed non-coding RNA was shown to modulate cell growth [48]. This lincRNA (TUC338) is predominantly located in the nucleus and has strongly increased expression in carcinoma cells compared to non-transformed hepatocytes. TUC338 is located in part within another gene (poly (rC) binding protein 2), but transcribed independently.

5 Implications for the Treatment of Cancer

Ultimately, the study of miRNAs and other non-coding RNAs can only make an impact on clinical hematology and oncology if new and better treatments for cancer can be developed. This goal can also be achieved by prognosticating new risk factors and better targeting the currently available treatments for cancer. Possibilities are to introduce tumor suppressor non-coding RNAs directly into cancer cells or to antagonize over-expressed cancer-promoting non-coding RNAs in cancer cells. Worthwhile topics of study are how resistance mechanisms can be overcome and how the currently available treatments for cancer (radiation, chemotherapy, cytokines, small molecules) interact with the non-coding RNAs in cancer cells. A potential advantage of gene therapy using miRNAs or similar non-coding molecules is that these molecules can be easily transfected because of their size. In addition, because one non-coding molecule regulates multiple genes, small changes in that molecule's expression *in vivo* may have a major impact on cancer signal transduction. Potential disadvantages of gene therapy may be lack of specificity and stability of the transfected molecules. Potential side effects also need careful attention.

The first step in testing a new treatment in humans is to study the drug or procedure in animal models, for example, in immunosuppressed (nude) mice. As an example, the transfection of myeloma cell lines with miR-19 and miR-181 antagonists resulted in significant tumor suppression in a xenograft mouse model [17]. In glioma, the transfection of the precursor of miR-34 into a glioma cell line led to a drastic reduction of tumor growth when injected into the brain of immunosuppressed mice [49].

Another miRNA that has been tested preclinically is miR-34a, which has reduced expression in several types of cancer, including lung cancer, and is considered a tumor suppressor miRNA. A group of scientists recently synthesized a miR-34a mimic and incorporated it into a lipid-based vector. When miR-34a was administered into tumors or into the systemic circulation in mice, the development of lung tumors was delayed or blocked. The authors showed that miR-34 accumulated in the tumors and its direct targets were downregulated. In this mouse model, few side effects were observed; in particular, no elevations of liver enzymes or cytokines occurred [50].

MiR-191 was recently identified as a potential target for gene therapy in hepatocellular carcinoma. In vitro, the inhibition of miR-191 decreased cell proliferation and induced apoptosis. In vivo, in an orthotopic xenograft mouse model, anti-miR-191 significantly reduced tumor masses. In addition, miR-191 was found to be upregulated by a known liver carcinogen (dioxin) and regulated various cancer-related pathways [51].

Especially in aggressive cancers, chemotherapy or radiation eradicates more than 98 % of tumor cells, but due to cancer stem cells, the cancer re-grows, develops metastasis and leads to death. Targeting miRNAs or other non-coding RNAs to cancer stem cells, alone or in combination with currently available treatments for cancer, would constitute a breakthrough in cancer therapy.

In disorders other than cancer (hepatitis, hypercholesterolemia), interesting preclinical models have been published. The miRNA miR-122 is expressed predominantly in liver cells and is essential for hepatitis C RNA replication. Chimpanzees chronically infected with hepatitis C were treated with modified oligonucleotides complementary to miR-122. This treatment led to a long-lasting decrease in the hepatitis C viral load (without increased resistance), accompanied by suppression of miR-122 [52]. At the same time, interferon-regulated genes were modulated.

In summary, the safe and effective administration of miRNAs and antagomirs in patients with cancer would have a major impact. Before this can happen, more work elucidating pathomechanisms and optimizing delivery of miRNAs and other non-coding RNAs is necessary. From the point of view of drug development, frequent cancers (such as lung cancer or breast cancer) and cancers for which no effective treatment is available for advanced stages (such as malignant melanoma or hepatocellular carcinoma) will have priority. Combining our growing understanding of non-coding RNA with the data from whole-genome sequencing, a clearer perspective of what causes cancer is on the horizon.

Acknowledgments G.A.C. is supported as a Fellow at The University of Texas MD Anderson Research Trust, as a Fellow of The University of Texas System Regents Research Scholar program and by the CLL Global Research Foundation. Work in Dr Calin's laboratory is supported in part by the National Institutes of Health (including MD Anderson's Cancer Center Support Grant, CA016672), by a Department of Defense Breast Cancer Idea Award, by Developmental Research Awards in Breast Cancer, Ovarian Cancer and Leukemia Specialized Programs of Research Excellence, by a CTT/3I-TD grant and by a 2009 Seena Magowitz—Pancreatic Cancer Action Network—AACR Pilot Grant. R.M. is supported by Louisiana State University, Shreveport (sabbatical leave). Sunita C. Patterson helped with editing.

References

1. Calin GA, Dumitru C, Shimizu M, et al (2002) Frequent deletions and down-regulation of micro-RNA genes miR15 and miR16 at 13q14 in chronic lymphocytic leukemia. Proc Natl Acad Sci USA 99:15524–15529
2. Lu J, Getz G, Miska EA, et al (2005) MicroRNA expression profiles classify human cancers. Nature 435:834–838
3. Volinia S, Calin GA, Liu CG, et al (2006) A microRNA expression signature of human solid tumors defines cancer gene targets. Proc Natl Acad Sci USA 103:2257–2261
4. Calin GA, Sevignani C, Dumitru CD, et al (2004) Human microRNA genes are frequently located at fragile sites and genomic regions involved in cancers. Proc Natl Acad Sci USA 101:2999–3004
5. Munker R, Calin GA (2011) MicroRNAs and Cancer. Encyclopedia of Life Sciences. doi: 10.1002/9780470015902.a0023161
6. Zhou H, Huang X, Cui H, et al (2010) miR-155 and its star form partner miR-155* cooperatively regulate type I interferon production by human plasmacytoid dendritic cells. Blood 116:5885–5894
7. Krol J, Loedige I, Filipowicz W (2010) The widespread regulation of MicroRNA biogenesis, function and decay. Nature Rev Genet 11:597–610
8. Calin GA, Dumitru C, Shimizu M, et al (2002) Frequent deletions and down-regulation of micro-RNA genes miR15 and miR16 at 13q14 in chronic lymphocytic leukemia. Proc Natl Acad Sci USA 99:15524–15529
9. Calin GA, Ferracin M, Cimmino A, et al (2005) MicroRNA signature associated with prognosis and progression in chronic lymphocytic leukemia. N Engl J Med 353:1793–1801
10. Scalione BJ, Salerno E, Balan M, et al (2007) Murine models of chronic lymphocytic leukaemia: role of microRNA-16 in the New Zealand black mouse model. Br J Haematol 139: 645–657
11. Klein U, Lia M, Crespo M, et al (2010) The DLEU2/miR-15a/16-1 cluster controls B cell proliferation and its deletion leads to chronic lymphocytic leukemia. Cancer Cell 17:28–40

12. Santanam U, Zanesi N, Efanov A, et al (2010) Chronic lymphocytic leukemia modeled in mouse by targeted miR-29 expression. Proc Natl Acad Sci USA 107:12210–12215
13. Ferracin M, Zagatti B, Rizotto L, et al (2010) MicroRNAs involvement in fludarabine refractory chronic lymphocytic leukemia. Molecular Cancer 9:123
14. Fabbri M, Bottoni A, Shimizu M, et al (2011) Association of a microRNA/TP53 feedback circuitry with the pathogenesis and outcome of B-chronic lymphocytic leukemia. J Am Med Assoc 305:59–67
15. Calin GA, Croce CM (2009) Chronic lymphocytic leukemia: interplay between noncoding RNAs and protein-coding genes. Blood 114:4761–4770
16. Lőffler D, Brocke-Heidrich K, Pfeifer G, et al (2007) Interleukin-6 dependent survival of multiple myeloma cells involves the Stat3-mediated induction of micro-RNA-21 through a highly-conserved enhancer. Blood 110:1330–1333
17. Pichiorri F, Su SS, Ladetto M, et al (2008) MicroRNAs regulate critical genes associated with multiple myeloma pathogenesis. Proc Natl Acad Sci USA 105:12885–12890
18. Pichiorri F, Su SS, Rocci A, et al (2010) Downregulation of p53-inducible microRNAs 192, 194, and 215 impairs the p53/MDM2 autoregulatory loop in multiple myeloma development. Cancer Cell 18:367–381
19. Lionetti L, Agnelli L, Mosca L, et al (2009) Integrative high-resolution microarray analysis of human myeloma cell lines reveals deregulated miRNA expression associated with allelic imbalances and gene expression profiles. Genes Chromosom Cancer 48:521–531
20. Todoerti K, Barbui V, Pedrini O, et al (2010) Pleiotropic antimyeloma activity of ITF2355; inhibition of interleukin-6 receptor signaling and repression of miR-19a and miR-19b. Haematologica 95:260–269
21. Roccaro AM, Sacco A, Thompson B, et al (2009) MicroRNAs 15a and 16 regulate tumor proliferation in multiple myeloma. Blood 113:6669–6680
22. Zhou Y, Chen L, Barlogie B, et al (2010) High-risk myeloma is associated with global elevation of miRNAs and overexpression of EIF2C2/AGO2. Proc Natl Acad Sci USA 107:7904–7909
23. Ballabio E, Mitchell T, van Kester MS, et al (2010) MicroRNA expression in Sézary syndrome: identification, function and diagnostic potential. Blood 116:1105–1113
24. Zhao JJ, Lin J, Lwin T, et al (2010) microRNA expression profile and identification of miR-29 as a prognostic marker and pathogenetic factor by targeting CDK6 in mantle cell lymphoma. Blood 115:2630–2639
25. Di Lisio L, Gómez-López G, Sanchez-Beato M, et al (2010) Mantle cell lymphoma: transcriptional regulation by microRNAs. Leukemia 24:1335–1342
26. Mitchell PS, Parkin RK, Kroh EM, et al (2008) Circulating microRNAs as stable blood-based markers for cancer detection. Proc Natl Acad Sci USA 105:10513–10518
27. Ng EKO, Chong WWS, Jin H, et al (2009) Differential expression of microRNAs in plasma of patients with colorectal cancer: a potential marker for colorectal cancer screening. Gut 58:1375–1381
28. Silva J, Garcia V, Zaballos Á, et al (2010) Vesicle-related microRNAs in plasma of NSCLC patients and correlation with survival. Eur Resp J Express E-pub. (July 2010) doi:10.1183/09031936.00029610
29. Li A, Omura N, Hong SM, et al (2010) Pancreatic cancers epigenetically silence SIP1 and hypomethylate and overexpress miR-200a/200b in association with elevated circulating miR-200a and miR-200b levels. Cancer Res 70:5226–5237
30. Heneghan HM, Miller N, Kelly R, et al (2010) Systemic miRNA-195 differentiates breast cancer from other malignancies and is a potential biomarker for detecting noninvasive and early stage disease. Oncologist 15:673–682
31. Weber JA, Baxter DH, Zhang S, et al (2010) The MicroRNA spectrum in 12 body fluids. Clin Chem 56:1733–1741
32. Malecová B, Morris KV (2010) Transcriptional gene silencing mediated by non-coding RNAs. Curr Opin Mol Ther 12:214–222

33. Calin GA, Liu CG, Ferracin M, et al (2007) Ultraconserved regions encoding ncRNAs are altered in human leukemias and carcinomas. Cancer Cell 12:215–229
34. Lujambio A, Portela A, Liz J, et al (2010) CpG island hypermethylation-associated silencing of non-coding RNAs transcribed from ultraconserved regions in human cancer. Oncog 29:6390–6401
35. Huarte M, Rinn JL (2010) Large non-coding RNAs: missing links in cancer? Hum Molec Genetics 19:R152–R161
36. Ji P, Diederichs S, Wang W, et al (2003) MALAT-1, a novel noncoding RNA, and thymosin β4 predict metastasis and survival in early-stage non-small cell lung cancer. Oncog 22:8031–8041
37. Lin R, Maeda S, Liu C, et al (2007) A large noncoding RNA is a marker for murine hepatocellular carcinomas and a spectrum of human carcinomas. Oncog 26:851–858
38. Fellenberg J, Barnd L, Delling G, et al (2007) Prognostic significance of drug-related genes in high-grade osteosarcoma. Modern Pathol 20:1085–1094
39. Tano K, Mizuno R, Okada T, et al (2010) MALAT-1 enhances cell motility of lung adenocarcinoma cells by influencing the expression of motility-related genes. FEBS Lett 584:4575–4580
40. Gupta RA, Shah N, Wang KC, et al (2010) Long non-coding RNA HOTAIR reprograms chromatin state to promote cancer metastasis. Nature 464:1071–1076
41. Yoshimizu T, Miroglio A, Ripoche MA, et al (2008) The H19 locus acts *in vivo* as a tumor suppressor. Proc Natl Acad Sci USA 105:12417–12422
42. Sirchia SM, Tabano S, Monti L, et al (2009) Misbehaviour of XIST RNA in breast cancer cells. PLoS 4:5559
43. Parrott AM, Tsai M, Batchu P, et al (2010) The evolution and expression of the snaR family of small non-coding RNAs. Nucl Ac Res E-pub 39:1485–1500
44. Mestdagh P, Fredlund E, Pattyn F, et al (2010) An integrative genomics screen uncovers ncRNA T-UCR functions in neuroblastoma tumours. Oncogene 29:3583–3592
45. Chung S, Nakagawa H, Uemura M, et al (2011) Association of a novel long non-coding RNA in 8q24 with prostate cancer susceptibility. Cancer Sci 102:245–252
46. Matouk IJ, Abbasi I, Hochberg A, et al (2009) Highly upregulated in liver cancer noncoding RNA is overexpressed in hepatic colorectal metastasis. Eur J Gastroenterol Hepatol 21:688–692
47. Romanuik TL, Wang G, Morozowa O, et al (2010) LNCaP atlas: gene expression associated with *in vivo* progression to castration-recurrent prostate cancer. BMC Medical Genomics 3:43
48. Braconi C, Valeri N, Kogure T, et al (2011) Expression and functional role of a transcribed noncoding RNA with an ultraconserved element in hepatocellular carcinoma. Proc Natl Acad Sci USA 108:786–791
49. Li Y, Guessous F, Zhang Y, et al (2009) MicroRNA-34a inhibits glioblastoma growth by targeting multiple oncogenes. Cancer Res 69:7569–7576
50. Wiggins JF, Ruffino L, Kelnar K, et al (2010) Development of a lung cancer therapeutic based on the tumor suppressor microRNA-34. Cancer Res 70:5923–5930
51. Elyakim E, Sitbon E, Faerman A, et al (2010) hsa-miR-191 is a candidate oncogene target for hepotocellular carcinoma therapy. Cancer Res 70:8077–8087
52. Lanford RE, Hildebrandt-Eriksen ES, Petri A, et al (2010) Therapeutic silencing of microRNA-122 in primates with chronic hepatitis C virus infection. Science 327:198–201

Function of miRNAs in Tumor Cell Proliferation

Zuoren Yu, Aydin Tozeren and Richard G. Pestell

Abstract MicroRNAs (miR) are a class of multifunctional, small, non-coding, singled-stranded molecules that regulate the stability or translational efficiency of targeted messenger RNAs. According to the miRBase Sequence Database (http://www.mirbase.org/index.shtml), more than 1,000 miR sequences have been identified from the tissues or cells of human origin. miRNAs are transcribed from the genome mostly by RNA polymerase II into primary miRNAs (called pri-miRNA) which are usually around 1 kb in length. pri-miRNAs are further processed in the nucleus by a ribonucleases complex composed of Drosha and DGCR8 into precursor miRNAs (called pre-miRNAs) which are around 70–90 nucleotides in length with imperfectly complementary stem-loop-stem structures. The pre-miRNA is then transported by exportin-5, a pre-miRNA-specific export carrier, to the cytoplasm where the pre-miRNA is cleaved by another ribonuclease, Dicer, into a double-stranded miRNA which consists of a mature miRNA sequence of about 17–25 nucleotides long and a miRNA* fragment (derived from the opposite strand to the mature miRNA strand). The mature miRNA is assembled into a ribonucleoprotein complex known as RNA-induced silencing complex (RISC) that includes Argonaute protein [1]. The miR-RISC complex could lead to base-pairing interactions between a miRNA and the binding site of its target mRNAs within the 3' untranslated region (3'UTR). The interaction could lead to endonucleotic cleavage of the target mRNA or interference with its ability to be translated depending on the base-pairing complementarity between the miRNA and the target mRNA [2, 3].

Z. Yu (✉)
Research Center for Translational Medicine,
Key Laboratory of Arrhythmia, East Hospital,
Tongji University School of Medicine, Shanghai 200120, China
e-mail: zuoren.yu@gmail.com

Aydin Tozeren
School of Biomedical Engineering, Science and Health Systems,
Drexel University, Philadelphia, PA 19104, USA

R. G. Pestell
Departments of Cancer Biology and Medical Oncology,
Kimmel Cancer Center, Thomas Jefferson University,
233 South, 10th Street, Philadelphia, PA 19107, USA
e-mail: richard.pestell@jefferson.edu

S. Alahari (ed.), *MicroRNA in Cancer,*
DOI 10.1007/978-94-007-4655-8_2,

1 Introduction

MicroRNAs (miR) are a class of multifunctional, small, non-coding, singled-stranded molecules that regulate the stability or translational efficiency of targeted messenger RNAs. According to the miRBase Sequence Database (http://www.mirbase.org/index.shtml), more than 1,000 miR sequences have been identified from the tissues or cells of human origin. miRNAs are transcribed from the genome mostly by RNA polymerase II into primary miRNAs (called pri-miRNA) which are usually around 1 kb in length. pri-miRNAs are further processed in the nucleus by a ribonucleases complex composed of Drosha and DGCR8 into precursor miRNAs (called pre-miRNAs) which are around 70–90 nucleotides in length with imperfectly complementary stem-loop-stem structures. The pre-miRNA is then transported by exportin-5, a pre-miRNA-specific export carrier, to the cytoplasm where the pre-miRNA is cleaved by another ribonuclease, Dicer, into a double-stranded miRNA which consists of a mature miRNA sequence of about 17–25 nucleotides long and a miRNA* fragment (derived from the opposite strand to the mature miRNA strand). The mature miRNA is assembled into a ribonucleoprotein complex known as RNA-induced silencing complex (RISC) that includes Argonaute protein [1]. The miR-RISC complex could lead to base-pairing interactions between a miRNA and the binding site of its target mRNAs within the 3′ untranslated region (3′UTR). The interaction could lead to endonucleotic cleavage of the target mRNA or interference with its ability to be translated depending on the base-pairing complementarity between the miRNA and the target mRNA [2, 3].

The target identification of miRNAs remains challenging due to the lack of a confident criteria or effective ways to predict targets accurately. Nevertheless, different bioinformatics approaches have been applied to search the putative targets for a particular miRNA. Usually the nucleotides 2–8 (called "seed" sequence) of a miRNA are considered as the most important sequence for binding to target mRNA. A perfect complementarity between target mRNA 3'UTR to the "seed" sequence of a miRNA is required for the target prediction. At this point each vertebrate miRNA is supposed to bind to as many as hundreds gene targets or even more. And each gene may contain multiple binding sites for different miRNAs. miRNAs have the potential to target about one-third of human mRNAs [4]. However, the regulatory interaction between any predicted target gene and a particular miRNA has to be experimentally confirmed, usually through luciferase reporter assays.

miRNAs have been demonstrated to regulate a broad range of biological processes including timing of development, cell cycle progression, embryonic stem cell, cancer stem cell, cancer initiation, cancer cell proliferation, cancer metastasis and apoptosis [5–11]. Cancer is caused by multiple processes including uncontrolled proliferation and the inappropriate survival of damaged cells. Many regulatory factors switch on or off genes that direct cellular proliferation and differentiation. Emerging evidence indicates miRNAs are involved in tumorigenesis and function as tumor suppressors or oncogenes [9]. Altered expression of miRNAs or mutations of miRNA genes have been described in different types of human cancer.

For example, let-7 is downregulated in several cancers including lung cancer [12]; miR-15a and miR16-1 are deleted and/or down-regulated in ~70 % of patients with chronic lymphocytic leukemia [13]; miR-17/20a are decreased in abundance in tumor sample from breast cancer patient compared to the matching sample from same patient [14]. The expression level of a particular miRNA varies by cell type. Thus the same miRNA may perform different functions through distinct pathways dependent on the tissue or cell type. It is important to understand the tumor cell type-specific pathway through which a miRNA regulates cancer cell proliferation and tumorigenesis.

2 Aberrant miRNA Expression in Cancer

miRNAs were linked to cancer very soon after their discovery. The first report linking miRNA to cancer patient was in 2002 by Calin et al. showing that miR-15 and miR-16 are located at chromosome 13q14, a region deleted in more than half of B cell chronic lymphocytic leukemias (B-CLL) [13]. Both of the miRNAs are deleted or down-regulated in approximately 68 % of CLL patients. Their further study mapping 186 miRNA locations in the genome indicated that 52.5 % of miRNA genes are located at cancer-associated genomic regions or in fragile sites [8]. In 2005, Lorio et al. identified 29 miRNAs with aberrant expression in human breast cancer by microarray and northern blot analyses on 76 breast tumor samples and 14 human breast cell lines [15]. Jiang et al. detected 222 human miRNA precursors in expression profile of 32 human cell lines from lung, breast, colorectal, hematologic, prostate, pancreatic, and head and neck cancers [16]. Several miRNAs had tissue-specific aberrant expression including miR-205 which showed 36-fold higher abundance in head and neck cancer cell lines than other cell lines. In 2006, Zhang and colleagues performed an analysis of 283 known human miRNA genes by array-based comparative genomic hybridization in 227 human ovarian cancer, breast cancer, and melanoma specimens demonstrating the high-frequency gene copy number abnormality of miRNA-containing regions throughout the genome in human ovarian cancer (37.1 %), breast cancer (72.8 %), and melanoma (85.9 %) [17]. Murakami et al. analyzed the miRNA expression profiles in 25 pairs of hepatocellular carcinoma (HCC), adjacent non-tumorous tissue (NT) and nine additional chronic hepatitis (CH) specimens using a human miRNA microarray [18]. Three miRNAs (miR-224, miR-18 and pre-miR-p18) exhibited higher expression and five miRNAs (miR-199a, miR-199a*, miR-200a, miR-125a and miR-195) showed lower expression in the HCC samples compared to the NT samples. Yanaihara et al. analyzed the miRNA expression in 104 pairs of primary lung cancers and corresponding non-cancerous lung tissues, and identified 43 miRNAs with differential expression in lung cancer [19]. Many of these miRNAs are located at frequently deleted or amplified regions in several malignancies. For example, miR-21 and miR-205 are located at the region amplified in lung cancer, whereas hsa-mir-126* and hsa-mir-126 are at a region deleted in lung cancer. Reduced expression of precursor let-7a and let-7f was also found in adenocarcinoma and squamous cell carcinoma. In 2007, Porkka et al. examined the miRNA expression profiling of six prostate cancer cell lines, nine prostate cancer xenografts samples, four benign prostatic hyperplasia, and nine prostate carcinoma samples using an oligonucleotide microarray [20]. They identified 51 miRNAs aberrantly expressed in prostate cancer including 37 miRNAs down-regulated and 14 miRNAs up-regulated in the prostate carcinoma samples. Gottardo et al. reported four human miRNAs (miR-28, miR-185, miR-27, and let-7f-2) significantly up-regulated in renal cell carcinoma compared to normal kidney, and ten human miRNAs (miR-223, miR-26b, miR-221, miR-103-1, miR-185, miR-23b, miR-203, miR-17–5p, miR-23a, and miR-205) up-regulated in bladder cancers compared to normal bladder mucosa [21]. In 2008, Schepeler et al. profiled the expression of 315 human miRNAs in ten normal mucosa samples and 49 stage II colon cancers using microarray technology [22]. Comparing with normal mucosa, 25 miRNAs were differentially expressed (7 down; 18 up) in microsatellite unstable colon cancers, and 54 miRNAs were differentially expressed (29 down; 25 up) in microsatellite stable colon cancers. miR-145 was identified as the lowest expression in colon cancer relative to normal tissue.

These studies provide evidence for new mechanisms by which aberrant expression of miRNAs and/or the loss or the gain of miRNA-containing genomic regions in a specific type of cancer may contribute to tumorigenesis. miRNAs may therefore serve as new diagnostic biomarkers and/or therapeutic tools for human cancers.

3 miRNA Regulation of Tumor Cell Division

Cancer is characterized by loss of cellular growth control, excess of cellular proliferation and altered cellular metabolism, invasion and metastasis. Understanding the mechanisms controlling cell division is important to developing novel anti-cancer therapies. Cyclins and cyclin-dependent kinases (CDKs) determine cell cycle progression. The mechanisms by which miRNAs regulate the cell cycle are increasingly well understood. Controlling cell-cycle represents new approaches to tumor cell inhibition.

3.1 Cell Division Cycle

Cyclin D1 serves as a cell cycle regulatory switch in actively proliferating cells. The *cyclin D1* gene encodes the regulatory subunit of the holoenzyme that phosphorylates and inactivates the pRb protein to promote G_1/S transition. Cyclin D1 binds to CDK4/6 forming the active complex, phosphorylates the retinoblastoma (Rb) resulting in the Rb dissociation from E2F complexes resulting in the transcriptional regulation of genes which contain E2F sites in their promoters such as cyclin E, cyclin A, DNA polymerase, thymidine kinase. Cyclin E binds to CDK2 forming the cyclin E-CDK2 complex, which pushes the cell cycle G_1/S transition. The cyclin A-CDK2 complex activates DNA synthesis. The cyclin B-cdc2 complex initiates the G_2/M transition. In contrary to the positive regulation of cell division cycle by cyclins and CDKs, two inhibitory families encode negative regulators of the cell cycle. The *cip/kip* family includes $p21^{CIP1}$, $p27^{KIP1}$and $p57^{KIP2}$ which arrest cell cycle at G_1 phase by inactivating cyclin-CDK complexes, and INK4a/ARF family including $p16^{INK4a}$ and $p14^{arf}$ [23].

Cyclin D1 is overexpressed in several cancer types including breast, esophageal and thyroid cancer, encoding a rate-limiting factor for proliferation of cancer cells in tissue culture [24–26]. Inhibition of *cyclin D1* expression *in vivo* suppressed breast cancer cell proliferation in nude nice [28]. Cyclin E is ovexpressed in 10 % of breast cancers. Emerging evidences have shown that miRNAs interact with *cyclins*, cyclin-dependent kinases (CDKs), *E2F*, *Rb* and CDK inhibitors thereby regulating cellular division and tumor growth [14, 27–30].

3.2 miRNAs Inhibiting Tumor Cell Proliferation

3.2.1 miR-15a and miR-16-1

miR-15a and miR-16-1 are deleted and/or down-regulated in chronic lymphocytic leukemia patients [13], prostate cancer [31] and pituitary tumors [32]. The anti-apoptotic gene BCL2 is negatively regulated by miR-15a and miR-16-1, leading to the inhibition of tumor growth. miR-15a and miR-16-1 induce cell cycle arrest at the G_1 phase by targeting cell cycle regulators including cyclin D1, cyclin E1, cyclin D3 and CDK6 [33, 34].

3.2.2 miR-17/20

miR-17/20 expression is decreased in human breast cancer specimen compared to the matching normal tissue suggesting a tumor suppressor function in breast cancer [14]. miR-17/20 binds to the cyclin D1 3'UTR in the MCF-7 breast cancer cells, inhibits the expression of cyclin D1, resulting in cell cycle arrest and suppression of cell proliferation [14, 35]. Besides cyclin D1, other cell-cycle related genes are regulated by miR-17/20 thereby controlling cell cycle progression [35].

3.2.3 miR-221/222

miR-221/222 is a miRNA cluster targeting the CDK inhibitors $p27^{KIP1}$ and $p57^{KIP2}$ [36]. miR-221/222 cluster ectopic expression decreases $p27^{KIP1}$ and $p57^{KIP2}$ abundance, activating cyclin E-CDK2 and cyclin A-CDK2 complexes, facilitating the G_1/S phase transition and DNA synthesis. This has been demonstrated in both human breast cancer and gastric cancer [30, 36].

3.2.4 Let-7

Let-7 is a tumor suppressor miRNA family. Let-7 family members are down-regulated in lung [12], colon [37], ovarian [38] and breast cancer [39]. Let-7 regulates tumorigenesis via *Ras, HMGA2, MYC* and/or *caspase-3* [39–43]. Let-7 overexpression inhibits tumor cell proliferation by targeting cyclin D1, cyclin D3, cyclin A, CDK 4 [44] and *CCNA2, CDC25 A,CDK6* and *CDK8* [45].

3.2.5 miR-29

miR-29 family (miR-29a, 29b, and 29c) have been associated with acute myelogeneous leukemia (AML), rhabdomyosarcoma, hepatocellular carcinoma and mantle cell lymphoma [46–49] by regulating cell apoptosis, cell cycle, and cell proliferation pathways. miR-29 overexpression induces apoptosis and inhibits tumor cell proliferation *in vitro*. In mantle cell lymphoma miR-29 inhibits CDK6 protein and mRNA levels. The down-regulation of miR-29 may cooperate with cyclin D1 in MCL pathogenesis [49].

3.2.6 miR-34, miR-192 and miR-215

miR-34, miR-192 and miR-215 regulate the p53 tumor suppressor network [50]. miR-34 overexpression arrests the cell cycle, induces apoptosis and inhibits cancer cell proliferation and colony formation by downregulating cyclin D1, cyclin E2, E2Fs and CDK4/6 [50–53]. miR-192 and miR-215 expression is reduced in colon cancer samples, and miR-192/ miR-215 suppresses colony formation and carcinogenesis via $p21^{CIP1}$ accumulation and cell cycle arrest [54]. $p21^{CIP1}$ accumulation is partially dependent on the presence of wild-type p53.

3.3 miRNAs Enhancing Tumor Cell Proliferation

3.3.1 miR-21

miR-21 overexpression has been observed frequently in a wide variety of cancers including breast cancer [55, 56], lung cancer [19] and liver cancer [57]. miR-21 induces MCF-7 cell-derived breast tumor growth in the xenograft mouse model [55]. In MCF-7 cells, miR-21 targets the antiapoptotic gene Bcl-2 [55], the tumor-suppressor gene *tropomyosin 1* (*TPM1*) [58] and the tumor suppressor protein Programmed Cell Death 4 (PDCD4) [56]. In human hepatocellular cancer cells miR-21 targets the tumorsuppressor gene PTEN, thereby enhancing tumor cell proliferation, migration, and invasion [57].

3.3.2 miR-27a

miR-27a expression is upregulated in kidney cancers [21] and breast cancer cell lines [59]. The zincfinger *ZBTB10* gene is a direct target of miR-27a [59]. InMDA-MB-231 human breast cancer cells, miR-27a inactivation induced *ZBTB10* expressionand reduced expression of oncogene *specificity proteins* (*Sp1*, *Sp3*, and *Sp4*) at the mRNA and protein levels [60]. The cyclin B-cdc2 complex inhibitor *Myt-1* is another target of miR-27a [60]. In MDA-MB-231cells miR-27a inhibits *Myt-1* expression increasing Cyclin B-cdc2 activity thereby promoting breast cancer cell proliferation.

3.3.3 miR-155

miR-155 is frequently up-regulated in breast cancer [61], lung cancer [19], pancreatic [62] and lymphomas [63]. In breast cancer, miR-155 expression induces cell survival, growth and chemoresistance by targeting the FOXO3a gene [61]. In pancreatic ductal adenocarcinoma (PDAC) cells, miR-155 is oncogenic by targeting the tumor suppressor gene *tumor protein 53-induced nuclear protein 1* (TP53INP1) [62].

4 miRNA Regulation of Cancer Stem Cells

Cancer stem cells (CSCs) are a subpopulation of stem-like cells within tumors. CSCs are characterized by their self-renewal capacity, an ability to differentiate into non-tumorigenic cell progeny, and their ability to seed tumors when transplanted into animal hosts [65]. CSCs have been demonstrated in several solid tumors including human breast cancer and brain cancer [64, 66], and melanoma, glioblastoma colon, pancreas, lung and prostate cancers. CSCs are isolated and enriched on the basis of cell surface markers (CD44, CD24 and/or CD133) dependent on tumor type. The leukaemic stem-like cells are fractioned by $CD34^{++}$ $CD38^{-}$ [67]; the mammary tumorigenic CSCs are isolated by $CD44^{+}$ $CD^{24-/low}$lineage^{-} [65]; the colon CSCs are isolated by $CD133^{+}$ [68]. Recently epithelial-specific antigen (ESA) and aldehyde dehydrogenase-1 (ALDH-1) were added into the candidate list of CSC-specific markers [69, 70, 71].

miRNAs regulate self-renewal and differentiation of ES cells, adult tissue stem cells, and CSCs. A subset of miRNAs (miR-142-3p, miR-451, miR-106a, miR-142-5p, miR-15b, miR-20a, miR-106b, miR-25 and miR-486) has altered expression in lung cancer progenitor cells [72]. Thirty seven miRNAs were deregulated in human breast cancer stem cells ($CD44^{+}$ $CD24^{-/low}$lineage^{-}) compared to the lineage^{-} nontumorigenic breast cancer cells [73]. Notably, members of miR-200 family including miR-200a, miR-200b and miR-200c, were down-regulated in the breast cancer stem cell population. The miR-200 family regulates epithelial to mesenchymal transition (EMT) in breast cancer [74]. Let-7 has a low expression in breast CSCs and

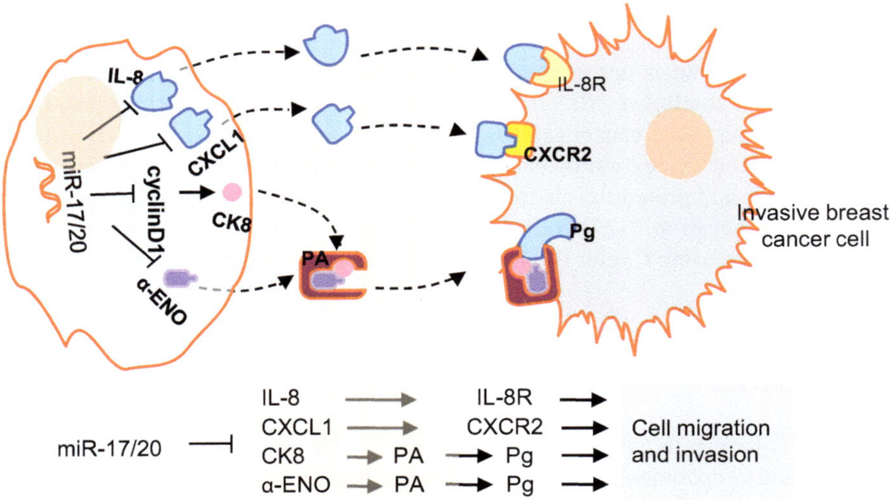

Fig. 1 miR-17/20 regulates cancer cell migration and invasion via heterotypic secreted signals *Pg* plasminogen, *PA* plasminogen activator, *IL-8R* IL-8 receptor, *CXCR2 CXCL1* receptor

increases with differentiation. Let-7 regulates self renewal and tumorigenicity of breast cancer stem cells [39]. Expression of let-7 in breast CSCs reduced proliferation, mammosphere formation, and the proportion of undifferentiated cells *in vitro* and tumor formation and metastasis in NOD/SCID mice [39].

5 miRNA Regulation of Tumor Microenvironment and Cancer Metastasis

The regulation of the tumor microenvironment to promote tumorigenesis was proposed by Paget in his 'soil and seed' hypothesis of cancer metastasis. Carcinogenesis and metastasis are controlled by both internal and external "heterotypic" signals from the surrounding cells and environment. Cancer metastases represent a complex process by which primary solid tumor cells invade adjacent tissue and grow into secondary tumors. miR-373 and miR-520c stimulate breast cancer cell migration and invasion by suppressing the gene *CD44* [75]; miRNA-200 and miR 205 inhibit EMT in breast cancer [74]. Our recent studies demonstrated a novel mechanism by which miRNA regulates cancer cell migration and invasion via heterotypic secreted signals (Fig. 1) [76]. miR-17/20 conditioned medium from cultures of MCF7 cells (a non-metastatic line) inhibited the migration and invasion of MDA-MB-231 cells (a metastatic line). miR-17/20 decreased the abundance of secreted factors such as cytokines (IL-8, CXCL1) and plasminogen activators (cytokeratin 8/18 and α-enolase) in MCF-7 cells. These secreted factors were essential for migration of the cancer cells. As miRNAs and the cancer microenvironment have a

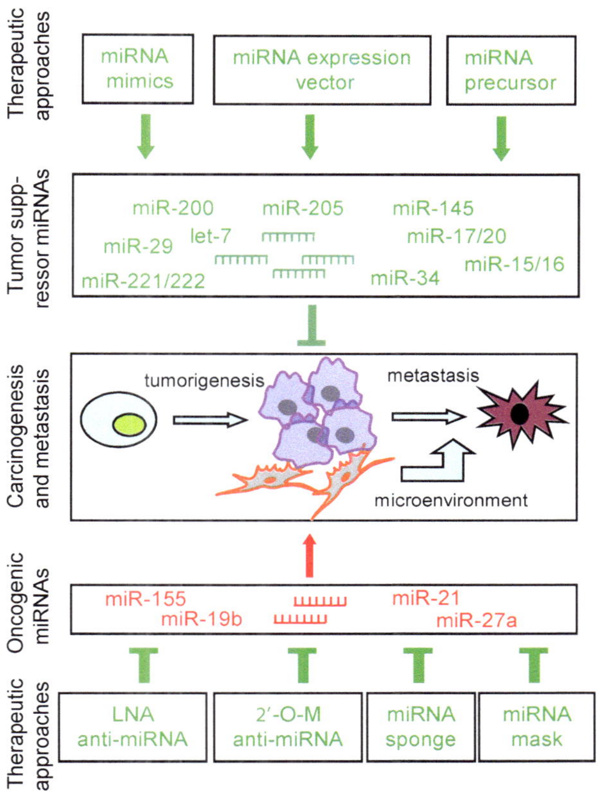

Fig. 2 miRNA-based cancer diagnostic and therapeutic strategies

crucial role in tumorigenesis and metastasis, these studies identify a potential new site of invention via inhibition of miRNA-regulated secreted factors.

6 Therapeutic Application of miRNA in Cancer

MiRNA-based cancer diagnostic and therapeutic approaches are being established and tested in animal models. There are two strategies for miRNA-based therapeutic application in cancers: expression-based restoration of tumor suppressor miRNA and functional inhibition of oncogenic miRNA (Fig. 2).

6.1 Restoring the Expression of Tumor Suppressor miRNAs

Two lines of evidence suggest miRNAs function as tumor suppressor rather than oncogenes. The global decrease of miRNA expression in cancer tissues compared to normal control [77], and the enhanced cellular transformation and tumorigenesis by impaired miRNA processing [78]. Synthetic miRNA mimics or miRNA expression vectors carrying either a pre-miRNA sequence or an artificial miRNA hairpin sequence have been widely used to perform the miRNA expression restoring *in vitro*. Recent publications demonstrated that miRNA reintroduction suppresses tumor growth *in vivo* [79, 80, 81]. Intranasal delivery of exogenous let-7 mimics to the lung tumors of mice (non-small-cell lung cancer) reduced the tumor burden [80]. Intravenous delivery of miR-34a mimics using a lipid-based delivery vehicle accumulated miR-34a in the tumor tissue in mouse models of non-small-cell lung cancer, and blocked tumor growth. Furthermore, this approach did not induce an immune response [81]. An adeno-associated virus (AAV)-mediated delivery of *miR-26a* to mouse liver resulted in reduced liver cancer cell proliferation, induction of tumor-specific apoptosis, and protection from disease progression without toxicity [82].

6.2 *Blocking the Function of Oncogenic miRNAs*

Since a small number of miRNAs show oncogenic function, such as miR-21, decreasing the expression level or blocking the function of those oncogenic miRNAs is another strategy for cancer therapy. Chemically modified antisense oligonucleotides (called anti-miRNA) are used most frequently to knock down miRNA *in vitro*. The modification includes addition of 2′-O-methyl, addition of 2′-O-methoxyethyl and locked nucleic acid (LNA) with 2′-O connecting to 4′-C. The modified nucleic acid structure has high affinity and high specificity to bind with target miRNA. Moreover, the anti-miRNA–miRNA structure is highly stable. As such the delivery of a specific anti-miRNA into cells prevents the miRNA from binding to their cognate target genes thereby silencing miRNA function.

Recently two more approaches were reported for blocking miRNAs. One is called the miRNA-sponge [83] which serves as a competitive inhibitor of miRNAs. An expression vector carrying multiple binding sites to a targeted miRNA is introduced into cells. Following the vector gene transcription, the over-expressed synthetic binding sequences occupy the endogenous miRNA in the cells with high affinity blocking miRNA regulation of its target genes. Another approach called the miRNA-mask [84] which uses oligonucleotides perfectly complementary to miRNA binding sites of target mRNAs. The miRNA-mask blocks the access of the miRNA to the binding sequence of the target mRNAs thereby blocking the miRNA-mRNA interaction.

Tumor-targeted delivery and local administration are still major challenges to apply miRNA therapy to the clinic. The possibility of immune response, off-target effects and toxicity of exogenous miRNA mimics or antagonism to normal tissues will have to be taken into account and minimized to a safe level.

Acknowledgments This work was supported in part by awards from National Institutes of Health [R01CA70896, R01CA75503, and R01CA86072 to R.G.P.]. Work conducted at the Kimmel Cancer Center was supported by the NIH Cancer Center Core grant [P30CA56036 to R.G.P.]. This project is supported by a generous grant from the Dr. Ralph and Marian C. Falk Medical Research Trust, and was funded and supported in part by a grant from the Pennsylvania Department of Health. The Department specifically disclaims responsibility for any analyses, interpretations or conclusions.

References

1. Meister G, Landthaler M, Patkaniowska A, Dorsett Y, Teng G, Tuschl T (2004) Human Argonaute2 mediates RNA cleavage targeted by miRNAs and siRNAs. Mol Cell 15:185–197
2. Ambros V (2004) The functions of animal microRNAs. Nature 431:350–355
3. Cullen BR (2004) Transcription and processing of human microRNA precursors. Mol Cell 16:861–865
4. Krek A, Grun D, Poy MN, Wolf R, Rosenberg L, Epstein EJ, MacMenamin P, da Piedade I, Gunsalus KC, Stoffel M, Rajewsky N (2005) Combinatorial microRNA target predictions. Nat Genet 37:495–500
5. Cui Q, Yu Z, Purisima EO, Wang E (2006) Principles of microRNA regulation of a human cellular signaling network. Mol Syst Biol 2:46
6. Tili E, Michaille JJ, Gandhi V, Plunkett W, Sampath D, Calin GA (2007) MiRNAs and their potential for use against cancer and other diseases. Future Oncol 3:521–537
7. Calin GA, Croce CM (2006) MicroRNA signatures in human cancers. Nat Rev Cancer 6:857–866
8. Calin GA, Sevignani C, Dumitru CD, Hyslop T, Noch E, Yendamuri S, Shimizu M, Rattan S, Bullrich F, Negrini M, Croce CM (2004) Human microRNA genes are frequently located at fragile sites and genomic regions involved in cancers. Proc Natl Acad Sci USA 101:2999–3004
9. Zhang B, Pan X, Cobb GP, Anderson TA (2007) MicroRNAs as oncogenes and tumor suppressors. Dev Biol 302:1–12
10. Esquela-Kerscher A, Slack FJ. (2006) Oncomirs—microRNAs with a role in cancer. Nat Rev Cancer 6:259–269
11. Cummins JM, Velculescu VE (2006) Implications of micro-RNA profiling for cancer diagnosis. Oncogene 25:6220–6227
12. Takamizawa J, Konishi H, Yanagisawa K, Tomida S, Osada H, Endoh H, Harano T, Yatabe Y, Nagino M, Nimura Y, Mitsudomi T, Takahashi T (2004) Reduced expression of the let-7 microRNAs in human lung cancers in association with shortened postoperative survival. Cancer Res 64:3753–3756
13. Calin GA, Dumitru CD, Shimizu M, Bichi R, Zupo S, Noch E, Aldler H, Rattan S, Keating M, Rai K, Rassenti L, Kipps T, Negrini M, Bullrich F, Croce CM (2002) Frequent deletions and down-regulation of micro-RNA genes miR15 and miR16 at 13q14 in chronic lymphocytic leukemia. Proc Natl Acad Sci USA 99:15524–15529
14. Yu Z, Wang C, Wang M, Li Z, Casimiro MC, Liu M, Wu K, Whittle J, Ju X, Hyslop T, McCue P, Pestell RG (2008) A cyclin D1/microRNA 17/20 regulatory feedback loop in control of breast cancer cell proliferation. J Cell Biol 182:509–517
15. Iorio MV, Ferracin M, Liu CG, Veronese A, Spizzo R, Sabbioni S, Magri E, Pedriali M, Fabbri M, Campiglio M, Menard S, Palazzo JP, Rosenberg A, Musiani P, Volinia S, Nenci I, Calin GA, Querzoli P, Negrini M, Croce CM (2005) MicroRNA gene expression deregulation in human breast cancer. Cancer Res 65:7065–7070
16. Jiang J, Lee EJ, Gusev Y, Schmittgen TD (2005) Real-time expression profiling of microRNA precursors in human cancer cell lines. Nucleic Acids Res 33:5394–5403

17. Zhang L, Huang J, Yang N, Greshock J, Megraw MS, Giannakakis A, Liang S, Naylor TL, Barchetti A, Ward MR, Yao G, Medina A, O'Brien-Jenkins A, Katsaros D, Hatzigeorgiou A, Gimotty PA, Weber BL, Coukos G (2006) microRNAs exhibit high frequency genomic alterations in human cancer. Proc Natl Acad Sci USA 103:9136–9141
18. Murakami Y, Yasuda T, Saigo K, Urashima T, Toyoda H, Okanoue T, Shimotohno K (2006) Comprehensive analysis of microRNA expression patterns in hepatocellular carcinoma and non-tumorous tissues. Oncogene 25:2537–2545
19. Yanaihara N, Caplen N, Bowman E, Seike M, Kumamoto K, Yi M, Stephens RM, Okamoto A, Yokota J, Tanaka T, Calin GA, Liu CG, Croce CM, Harris CC (2006) Unique microRNA molecular profiles in lung cancer diagnosis and prognosis. Cancer Cell 9:189–198
20. Porkka KP, Pfeiffer MJ, Waltering KK, Vessella RL, Tammela TL, Visakorpi T (2007) MicroRNA expression profiling in prostate cancer. Cancer Res 67:6130–6135
21. Gottardo F, Liu CG, Ferracin M, Calin GA, Fassan M, Bassi P, Sevignani C, Byrne D, Negrini M, Pagano F, Gomella LG, Croce CM, Baffa R (2007) Micro-RNA profiling in kidney and bladder cancers. Urol Oncol 25:387–392
22. Schepeler T, Reinert JT, Ostenfeld MS, Christensen LL, Silahtaroglu AN, Dyrskjot L, Wiuf C, Sorensen FJ, Kruhoffer M, Laurberg S, Kauppinen S, Orntoft TF, Andersen CL (2008) Diagnostic and prognostic microRNAs in stage II colon cancer. Cancer Res 68:6416–6424
23. Pestell RG, Albanese C, Reutens AT, Segall JE, Lee RJ, Arnold A (1999) The cyclins and cyclin-dependent kinase inhibitors in harmonal regulation of prolifertion and differentiation. Endocr Rev Aug 20(4):501–534
24. Fu M, Wang C, Li Z, Sakamaki T, Pestell RG (2004) Minireview: Cyclin D1: Normal and abnormal functions. Endocrinology 145:5439–5447
25. Jiang W, Kahn SM, Tomita N, Zhang YJ, Lu SH, Weinstein IB (1992) Amplification and expression of the human cyclin D gene in esophageal cancer. Cancer Res 52:2980–2983
26. Lazzereschi D, Sambuco L, Carnovale Scalzo C, Ranieri A, Mincione G, Nardi F, Colletta G (1998) Cyclin D1 and Cyclin E expression in malignant thyroid cells and in human thyroid carcinomas. Int J Cancer 76:806–811
27. O'Donnell KA, Wentzel EA, Zeller KI, Dang CV, Mendell JT (2005) C-Myc-regulated microRNAs modulate E2F1 expression. Nature 435:839–843
28. Lee RJ, Albanese C, Fu M, D'Amico M, Lin B, Watanabe G, Haines GK 3rd, Siegel PM, Hung MC, Yarden Y, Horowitz JM, Muller WJ, Pestell RG (2000) Cyclin D1 is required for transformation by activated Neu and is induced through an E2F-dependent signaling pathway. Mol Cell Biol Jan 20(2):672–683
29. Lu Y, Thomson JM, Wong HY, Hammond SM, Hogan BL (2007) Transgenic over-expression of the microRNA miR-17-92 cluster promotes proliferation and inhibits differentiation of lung epithelial progenitor cells. Dev Biol 310:442–453
30. Miller TE, Ghoshal K, Ramaswamy B, Roy S, Datta J, Shapiro CL, Jacob S, Majumder S (2008) MicroRNA-221/222 confers tamoxifen resistance in breast cancer by targeting p27Kip1. J Biol Chem 283:29897–29903
31. Bonci D, Coppola V, Musumeci M, Addario A, Giuffrida R, Memeo L, D'Urso L, Pagliuca A, Biffoni M, Labbaye C, Bartucci M, Muto G, Peschle C, De Maria R (2008) The miR-15a-miR-16-1 cluster controls prostate cancer by targeting multiple oncogenic activities. Nat Med 14:1271–1277
32. Bottoni A, Piccin D, Tagliati F, Luchin A, Zatelli MC, degli Uberti EC (2005) MiR-15a and miR-16-1 down-regulation in pituitary adenomas. J Cell Physiol 204:280–285
33. Deshpande A, Pastore A, Deshpande AJ, Zimmermann Y, Hutter G, Weinkauf M, Buske C, Hiddemann W, Dreyling M (2009) 3'UTR mediated regulation of the cyclin D1 proto-oncogene. Cell Cycle 8:3584–3592
34. Liu Q, Fu H, Sun F, Zhang H, Tie Y, Zhu J, Xing R, Sun Z, Zheng X (2008) MiR-16 family induces cell cycle arrest by regulating multiple cell cycle genes. Nucleic Acids Res 36:5391–5404
35. Yu Z, Baserga R, Chen L, Wang C, Lisanti MP, Pestell RG (2010) MicroRNA, cell cycle, and human breast cancer. Am J Pathol 176:1058–1064

36. Kim YK, Yu J, Han TS, Park SY, Namkoong B, Kim DH, Hur K, Yoo MW, Lee HJ, Yang HK, Kim VN (2009) Functional links between clustered microRNAs: suppression of cell-cycle inhibitors by microRNA clusters in gastric cancer. Nucleic Acids Res 37:1672–1681
37. Schetter AJ, Leung SY, Sohn JJ, Zanetti KA, Bowman ED, Yanaihara N, Yuen ST, Chan TL, Kwong DL, Au GK, Liu CG, Calin GA, Croce CM, Harris CC (2008) MicroRNA expression profiles associated with prognosis and therapeutic outcome in colon adenocarcinoma. JAMA 299:425–436
38. Yang N, Kaur S, Volinia S, Greshock J, Lassus H, Hasegawa K, Liang S, Leminen A, Deng S, Smith L, Johnstone CN, Chen XM, Liu CG, Huang Q, Katsaros D, Calin GA, Weber BL, Butzow R, Croce CM, Coukos G, Zhang L (2008) MicroRNA microarray identifies Let-7i as a novel biomarker and therapeutic target in human epithelial ovarian cancer. Cancer Res 68:10307–10314
39. Yu F, Yao H, Zhu P, Zhang X, Pan Q, Gong C, Huang Y, Hu X, Su F, Lieberman J, Song E (2007) let-7 regulates self renewal and tumorigenicity of breast cancer cells. Cell 131: 1109–1123
40. Mayr C, Hemann MT, Bartel DP. (2007) Disrupting the pairing between let-7 and Hmga2 enhances oncogenic transformation. Science 315:1576–1579
41. Johnson SM, Grosshans H, Shingara J, Byrom M, Jarvis R, Cheng A, Labourier E, Reinert KL, Brown D, Slack FJ (2005) RAS is regulated by the let-7 microRNA family. Cell 120:635–647
42. Tsang WP, Kwok TT (2008) Let-7a microRNA suppresses therapeutics-induced cancer cell death by targeting caspase-3. Apoptosis 13:1215–1222
43. Sampson VB, Rong NH, Han J, Yang Q, Aris V, Soteropoulos P, Petrelli NJ, Dunn SP, Krueger LJ (2007) MicroRNA let-7a down-regulates MYC and reverts MYC-induced growth in Burkitt lymphoma cells. Cancer Res 67:9762–9770
44. Schultz J, Lorenz P, Gross G, Ibrahim S, Kunz M (2008) MicroRNA let-7b targets important cell cycle molecules in malignant melanoma cells and interferes with anchorage-independent growth. Cell Res 18:549–557
45. Johnson CD, Esquela-Kerscher A, Stefani G, Byrom M, Kelnar K, Ovcharenko D, Wilson M, Wang X, Shelton J, Shingara J, Chin L, Brown D, Slack FJ (2007) The let-7 microRNA represses cell proliferation pathways in human cells. Cancer Res 67:7713–7722
46. Garzon R, Heaphy CE, Havelange V, Fabbri M, Volinia S, Tsao T, Zanesi N, Kornblau SM, Marcucci G, Calin GA, Andreeff M, Croce CM (2009) MicroRNA 29b functions in acute myeloid leukemia. Blood 114:5331–5341
47. Wang H, Garzon R, Sun H, Ladner KJ, Singh R, Dahlman J, Cheng A, Hall BM, Qualman SJ, Chandler DS, Croce CM, Guttridge DC (2008) NF-kappaB-YY1-miR-29 regulatory circuitry in skeletal myogenesis and rhabdomyosarcoma. Cancer Cell 14:369–381
48. Xiong Y, Fang JH, Yun JP, Yang J, Zhang Y, Jia WH, Zhuang SM (2010) Effects of microRNA-29 on apoptosis, tumorigenicity, and prognosis of hepatocellular carcinoma. Hepatol 51:836–845
49. Zhao JJ, Lin J, Lwin T, Yang H, Guo J, Kong W, Dessureault S, Moscinski LC, Rezania D, Dalton WS, Sotomayor E, Tao J, Cheng JQ (2010) microRNA expression profile and identification of miR-29 as a prognostic marker and pathogenetic factor by targeting CDK6 in mantle cell lymphoma. Blood 115:2630–2639
50. He L, He X, Lim LP, de Stanchina E, Xuan Z, Liang Y, Xue W, Zender L, Magnus J, Ridzon D, Jackson AL, Linsley PS, Chen C, Lowe SW, Cleary MA, Hannon GJ (2007) A microRNA component of the p53 tumour suppressor network. Nature 447:1130–1134
51. Tarasov V, Jung P, Verdoodt B, Lodygin D, Epanchintsev A, Menssen A, Meister G, Hermeking H (2007) Differential regulation of microRNAs by p53 revealed by massively parallel sequencing: miR-34a is a p53 target that induces apoptosis and G1-arrest. Cell Cycle 6:1586–1593
52. Sun F, Fu H, Liu Q, Tie Y, Zhu J, Xing R, Sun Z, Zheng X (2008) Downregulation of CCND1 and CDK6 by miR-34a induces cell cycle arrest. FEBS Lett 582:1564–1568

53. Corney DC, Flesken-Nikitin A, Godwin AK, Wang W, Nikitin AY (2007) MicroRNA-34b and MicroRNA-34c are targets of p53 and cooperate in control of cell proliferation and adhesion-independent growth. Cancer Res 67:8433–8438
54. Braun CJ, Zhang X, Savelyeva I, Wolff S, Moll UM, Schepeler T, Orntoft TF, Andersen CL, Dobbelstein M (2008) p53-Responsive micrornas 192 and 215 are capable of inducing cell cycle arrest. Cancer Res 68:10094–10104
55. Si ML, Zhu S, Wu H, Lu Z, Wu F, Mo YY (2007) miR-21-mediated tumor growth. Oncogene 26:2799–2803
56. Frankel LB, Christoffersen NR, Jacobsen A, Lindow M, Krogh A, Lund AH (2008) Programmed cell death 4 (PDCD4) is an important functional target of the microRNA miR-21 in breast cancer cells. J Biol Chem 283:1026–1033
57. Meng F, Henson R, Wehbe-Janek H, Ghoshal K, Jacob ST, Patel T (2007) MicroRNA-21 regulates expression of the PTEN tumor suppressor gene in human hepatocellular cancer. Gastroenterology 133:647–658
58. Zhu S, Si ML, Wu H, Mo YY (2007) MicroRNA-21 targets the tumor suppressor gene tropomyosin 1 (TPM1). J Biol Chem 282:14328–14336
59. Scott GK, Mattie MD, Berger CE, Benz SC, Benz CC (2006) Rapid alteration of microRNA levels by histone deacetylase inhibition. Cancer Res 66:1277–1281
60. Mertens-Talcott SU, Chintharlapalli S, Li X, Safe S (2007) The oncogenic microRNA-27a targets genes that regulate specificity protein transcription factors and the G2-M checkpoint in MDA-MB-231 breast cancer cells. Cancer Res 67:11001–11011
61. Kong W, He L, Coppola M, Guo J, Esposito NN, Coppola D, Cheng JQ (2010) MicroRNA-155 regulates cell survival, growth, and chemosensitivity by targeting FOXO3a in breast cancer. J Biol Chem 285:17869–17879
62. Metzler M, Wilda M, Busch K, Viehmann S, Borkhardt A (2004) High expression of precursor microRNA-155/BIC RNA in children with Burkitt lymphoma. Genes Chromosomes Cancer 39:167–169
63. Gironella M, Seux M, Xie MJ, Cano C, Tomasini R, Gommeaux J, Garcia S, Nowak J, Yeung ML, Jeang KT, Chaix A, Fazli L, Motoo Y, Wang Q, Rocchi P, Russo A, Gleave M, Dagorn JC, Iovanna JL, Carrier A, Pebusque MJ, Dusetti NJ (2007) Tumor protein 53-induced nuclear protein 1 expression is repressed by miR-155, and its restoration inhibits pancreatic tumor development. Proc Natl Acad Sci USA 104:16170–16175
64. Al-Hajj M, Wicha MS, Benito-Hernandez A, Morrison SJ, Clarke MF (2003) Prospective identification of tumorigenic breast cancer cells. Proc Natl Acad Sci USA 100:3983–3988
65. Gupta PB, Chaffer CL, et al (2009) Cancer stem cells: mirage or reality? Nature Med 15:1010–1012
66. Singh SK, Clarke ID, Terasaki M, Bonn VE, Hawkins C, Squire J, Dirks PB (2003) Identification of a cancer stem cell in human brain tumors. Cancer Res 63:5821–5828
67. Lapidot T, Sirard C, Vormoor J, Murdoch B, Hoang T, Caceres-Cortes J, Minden M, Paterson B, Caligiuri MA, Dick JE (1994) A cell initiating human acute myeloid leukaemia after transplantation into SCID mice. Nature 367:645–648
68. Ricci-Vitiani L, Lombardi DG, Pilozzi E, Biffoni M, Todaro M, Peschle C, De Maria R (2007) Identification and expansion of human colon-cancer-initiating cells. Nature 445: 111–115
69. Li C, Heidt DG, Dalerba P, Burant CF, Zhang L, Adsay V, Wicha M, Clarke MF, Simeone DM (2007) Identification of pancreatic cancer stem cells. Cancer Res 67:1030–1037
70. Ginestier C, Hur MH, Charafe-Jauffret E, Monville F, Dutcher J, Brown M, Jacquemier J, Viens P, Kleer CG, Liu S, Schott A, Hayes D, Birnbaum D, Wicha MS, Dontu G (2007) ALDH1 is a marker of normal and malignant human mammary stem cells and a predictor of poor clinical outcome. Cell Stem Cell 1:555–567
71. Velasco-Velázquez MA, Yu Z, Jiao X, et al (2009) Cancer stem cells and the cell cycle: targeting the drive behind breast cancer. Expert Rev Anticancer Ther 9:275–9

72. Qian S, Ding JY, Xie R, An JH, Ao XJ, Zhao ZG, Sun JG, Duan YZ, Chen ZT, Zhu B (2008) MicroRNA expression profile of bronchioalveolar stem cells from mouse lung. Biochem Biophys Res Commun 377:668–673
73. Shimono Y, Zabala M, Cho RW, Lobo N, Dalerba P, Qian D, Diehn M, Liu H, Panula SP, Chiao E, Dirbas FM, Somlo G, Pera RA, Lao K, Clarke MF (2009) Downregulation of miRNA-200c links breast cancer stem cells with normal stem cells. Cell 138:592–603
74. Gregory PA, Bert AG, Paterson EL, Barry SC, Tsykin A, Farshid G, Vadas MA, Khew-Goodall Y, Goodall GJ (2008) The miR-200 family and miR-205 regulate epithelial to mesenchymal transition by targeting ZEB1 and SIP1. Nat Cell Biol 10:593–601
75. Huang Q, Gumireddy K, Schrier M, le Sage C, Nagel R, Nair S, Egan DA, Li A, Huang G, Klein-Szanto AJ, Gimotty PA, Katsaros D, Coukos G, Zhang L, Pure E, Agami R (2008) The microRNAs miR-373 and miR-520c promote tumour invasion and metastasis. Nat Cell Biol 10:202–210
76. Yu Z, Willmarth NE, Zhou J, Katiyar S, Wang M, Liu Y, McCue PA, Quong AA, Lisanti MP, Pestell RG (2010) MicroRNA 17/20 inhibits cellular invasion and tumor metastasis in breast cancer by heterotypic signaling. Proc Natl Acad Sci USA 107:8231–8236
77. Lu J, Getz G, Miska EA, Alvarez-Saavedra E, Lamb J, Peck D, Sweet-Cordero A, Ebert BL, Mak RH, Ferrando AA, Downing JR, Jacks T, Horvitz HR, Golub TR (2005) MicroRNA expression profiles classify human cancers. Nature 435:834–838
78. Kumar MS, Lu J, Mercer KL, Golub TR, Jacks T (2007) Impaired microRNA processing enhances cellular transformation and tumorigenesis. Nat Genet 39:673–677
79. Bader AG, Brown D, Winkler M (2010) The promise of microRNA replacement therapy. Cancer Res 70:7027–7030
80. Trang P, Medina PP, Wiggins JF, Ruffino L, Kelnar K, Omotola M, Homer R, Brown D, Bader AG, Weidhaas JB, Slack FJ (2010) Regression of murine lung tumors by the let-7 microRNA. Oncogene 29:1580–1587
81. Wiggins JF, Ruffino L, Kelnar K, Omotola M, Patrawala L, Brown D, Bader AG (2010) Development of a lung cancer therapeutic based on the tumor suppressor microRNA-34. Cancer Res 70:5923–5930
82. Kota J, Chivukula RR, O'Donnell KA, Wentzel EA, Montgomery CL, Hwang HW, Chang TC, Vivekanandan P, Torbenson M, Clark KR, Mendell JR, Mendell JT (2009) Therapeutic microRNA delivery suppresses tumorigenesis in a murine liver cancer model. Cell 137:1005–1017
83. Ebert MS, Neilson JR, Sharp PA (2007) MicroRNA sponges: competitive inhibitors of small RNAs in mammalian cells. Nat Methods 4:721–726
84. Choi WY, Giraldez AJ, Schier AF (2007) Target protectors reveal dampening and balancing of Nodal agonist and antagonist by miR-430. Science 318:271–274

MicroRNAs in Cancer Stem Cells

Alexander Swarbrick

Abstract This chapter will address the emerging role for microRNAs in the control of cancer stem cell (CSC) biology. The concept of cancer stem cells is revolutionising our understanding of cancer cell biology. There is increasing evidence for a central role of microRNAs in control of the normal and malignant stem cell phenotype. This new understanding promises to open new avenues in cancer prognosis and therapy.

1 The Cancer Stem Cell Hypothesis

The cancer stem cell (CSC) hypothesis posits that the growth and maintenance of tumors is driven by a small proportion of cells. These cells are enriched for self renewal capacity and give rise to more differentiated progeny that make up the bulk of the tumor. Self renewal is distinct from other forms of proliferation in that at last one of the progeny is identical to its parent. Thus tumor heterogeneity is generated by the asymmetric division of CSCs and differentiation of their progeny, resembling the processes of lineage specification and differentiation observed in normal tissues. When transplanted to naive hosts, CSCs are capable of regenerating tumors displaying the same features and heterogeneity as their source cancer.

The CSC hypothesis challenges the notion that cancers are homogenous and raises significant issues in the way we treat cancer. It suggests that to eradicate a cancer we must target the CSC, as well as the bulk of the tumor. There is evidence for innate resistance of CSCs to commonly used therapies by virtue of the overexpression of drug efflux pumps, and other mechanisms. It follows then that chemotherapy may kill the bulk of the cancer but leave the CSC pool intact, thus remission following clinical response to therapy may be driven by this small pool of CSC that escaped therapy. There is also accumulating evidence for CSCs in metastasis,

---

A. Swarbrick (✉)
Garvan Institute of Medical Research,
384 Victoria st, Darlinghurst, NSW 2010, Australia
e-mail: a.swarbrick@garvan.org.au

S. Alahari (ed.), *MicroRNA in Cancer,*
DOI 10.1007/978-94-007-4655-8_3,

since metastatic dissemination requires extensive proliferative capacity and CSCs are proposed by some to be inherently motile and invasive [1, 10].

While the existence of CSCs in blood cancers such as leukaemia is relatively well accepted [38], the generality of the CSC hypothesis to other cancer types is widely debated. An alternative hypothesis, commonly known as the 'clonal evolution' hypothesis sees cancer cells evolving independently, generating tumor heterogeneity through non-hierarchical evolution driven by cell autonomous mutations, epigenetic changes and interactions with the surrounding tumor stroma [40, 52]. Of course, this concept is not incompatible with elements of the CSC hypothesis. One can imagine a hybrid model whereby tumor heterogeneity is driven by the integration of vestigial developmental processes, overlaid by epigenetic dynamics, interaction with the microenvironment and aberrant oncogenic signalling driven by mutation. For this reason, many in the field use the operational term 'tumor initiating cell' or 'tumor propagating cell', which simply define a cell capable of malignant propagation, irrespective of its source or position within a cellular hierarchy. This review will refer to CSCs, but acknowledges that the mechanisms responsible for generating cells with the features described above are still unclear, in particular that there is little evidence for a stem cell origin for CSCs.

The gold standard for experimental determination of cancer stem cell frequency relies on the passage of tumor cells into mice. Different doses of cells are transplanted and the proportion of mice developing tumors at each cell dose is scored to calculate the frequency of CSCs in the original population. Tumors are also phenotyped, to determine whether they resemble the original tumor, as would be required of a cancer repopulated by a CSC. To perform these experiments, tumors go through extensive handling and processing, which all have the potential to alter the behaviour of the cancers. Following surgical resection, tumors are diced, disaggregated, filtered, then frequently stained with antibodies and sorted by Flow cytometry, based on cell surface antigen expression. Cells are finally transplanted to recipients, which must be immuno-deficient to permit growth of human cells. Partly because of the difficulties and complexities of these experimental approaches, there is considerable debate around the existence of cancer stem cells. For example, in a series of manuscripts in Nature, different groups report vastly different proportions of cancer stem cells from human melanomas. While initial reports suggested that only ~1 in every million cells from melanomas are tumorigeneic in SCID mice [57], Quitana et. al. later demonstrated that simple modifications in protocol, including the use of more severely immunodeficient mice, is permissive for tumor growth by ~1/4 of unselected melanoma cells [53], suggesting that rare cancer stem cells do not exist in melanoma. More recently, Boiko et al revisit the issue and show that prospective isolation of CD271-positive cells from melanomas can significantly enrich for cancer stem cells able to form metastatic tumors in mice [2], suggesting the existence of a tumor-initiating hierarchy. Such disparate results within the one disease highlight the many problems with interpreting results from the field.

2 Links between EMT and the CSC phenotype

The epithelial mesenchymal transition (EMT) is a developmental process whereby epithelial cells lose cell-cell and cell-matrix adhesions. Markers of epithelial identity such as E-cadherin are replaced by the expression of mesenchymal markers such as vimentin (reviewed in [67]). The EMT permits cellular migration and increasing evidence is implicating EMT in metastatic progression, as a way for epithelial cancers to become motile and spread to distant organs. A central role for a number of microRNAs in the EMT, in particular members of the miR-200 family, has recently been reported by several groups [19, 51, 72] and this is the subject of another chapter in this book.

Of relevance to CSCs, Mani et al demonstrate that experimental induction of an EMT, by treatment with TGF-β or by expression of the transcription factor Twist, is sufficient to drive the acquisition of stem- or progenitor-like traits in immortalised mammary epithelial cells [43]. This work was subsequently confirmed by other groups [35, 47]. However, it is important to note that these cells are immortalised by the expression of Viral Large T antigen and the h-Tert subunit of telomerase [13] which regulate the activity of many key genes controlling self renewal and proliferative capacity, including p53, Rb and PP2A. TGF-β treatment of normal cells typically elicits strongly anti-proliferative and/or apoptotic responses, and TGF-β signalling components are commonly lost during neoplastic transformation, demonstrating that TGF-β signalling is tumor suppressive during early stages of tumorigenesis [7]. Therefore, whether an EMT is sufficient to drive the acquisition of the stem cell phenotype in truly normal cells remains unclear. However, the evidence that EMT can drive the maintenance of the CSC phenotype in transformed cells appears stronger and this is consistent with the acquisition of high levels of TGF-β signalling in advanced cancers [59]. There are, however, intriguing exceptions that challenge the universality of this mechanism. For instance, overexpression of miR-200, which is proposed to be a key repressor of the EMT and therefore would be predicted to repress metastatic dissemination, promotes metastatic dissemination in certain models, such as allografts of the commonly used 4TO7 and 4T1 isogenic mouse mammary carcinoma cells [12]. Furthermore, high expression of Twist, a transcription factor frequently shown to promote the EMT, is associated with differentiation of epithelial ovarian CSC to non-CSC progeny, via positive transcriptional regulation of miR-199a and miR-204 [74].

3 microRNAs in CSCs

Erwei Song and colleaugues have contributed significantly to our understanding of the role for microRNAs in CSCs. Song et. al. have shown that a number of microRNAs are specifically downregulated in breast cancer cell lines enriched for CSC activity. These include the Let-7 family of microRNAs, miR-30 and miR-200 family members [76]. In this context, Let-7 functions to repress self renewal and promote differentiation of breast cancer cell lines [76] and overexpression of Let-7 reduced tumor formation *in vivo*. The Ras proto-oncogene appears to be a critical Let-7 target controlling self renewal of cells, while targeting of HMG2A by Let-7 promotes differentiation [16]. miR-30 appears to have a similar function to Let-7 in repressing the self renewal capacity of normal and malignant mammary epithelium [75]. Overexpression of miR-30 is accompanied by a reduction in the mammosphere-forming capacity of breast cancer cells and downregulation of Integrin beta-3 and Ubc9. shRNA-mediated knockdown of these targets phenocopies mIR-30 overexpression. One of the caveats of the work identifying a role for miR-30 and Let-7 in CSC [76] is that chronic chemotherapy treatment was used to enrich for CSCs, on the basis that CSCs are thought to possess innate resistance to chemotherapy [60]. However, it is likely that the pressure of chemotherapy treatment would lead to changes in microRNA expression that are not related to the cancer stem cell phenotype in untreated cells.

Interestingly, Let-7 microRNAs are also differentially expressed in tissue stem and progenitor cells compared to their differentiated progeny and can control stem and progenitor cell phenotype. Ibarra et. al reported low levels of the Let-7 microRNAs in a mouse mammary cell line that contains stem and progenitor cells [26]. Interestingly, these authors use a Let-7 'sensor', which reports on the cellular activity of Let-7, to show that prospective isolation of cells with low Let-7 activity identifies the population of cells with high clonogenic potential.

Instead of using cytotoxic therapy to enrich for CSCs, Yohei Shimono in Michael Clarke's group analysed populations of $CD24^{-}$ $CD44^{+}$ human breast CSCs isolated by FACS and identify a number of microRNAs differentially expressed in CSCs [58]. Three clusters of microRNAs from the miR-200 seed sequence family, miR-200c-141, miR-200b-200a-429, and miR-183–96-182, were downregulated in human breast CSCs and in normal human and mouse mammary stem cells. Expression of mIR-200c was sufficient to suppress normal stem cell activity in a mammary reconstitution assay, and cancer cell proliferation and tumorigenicity in transplantation assays. Ulrich Wellner and colleagues [72] similarly show that members of the miR-200 family of microRNAs, miR-183 and miR-203 were under the control of the ZEB transcription factor and act to repress mouse ES cell self renewal as well as tumorsphere formation by cancer cells. Interestingly, Let-7 and mIR-200 family members are poorly expressed by chemotherapy-resistant cancer cells, concomitant with the appearance of a mesenchymal phenotype [39], both of which are traits associated with the CSC phenotype. These observations provide further tantalising evidence for a connection between microRNAs, EMT and the CSC phenotype.

Both of the above studies [58, 72] identify Bmi1 as a key target of these microRNAs in controlling self renewal. Bmi1 is a member of the polycomb complex, responsible for transcriptional repression of genes by chromatin modification [54]. Bmi1 is required for stem cell function in multiple normal stem cells [50], and has been previously implicated in the control of breast CSC function [41]. Bmi1 is emerging as a central point of control in CSC biology, and miR-128 has recently been shown to target Bmi1 in glioma brain tumors [18]. miR-128 is poorly expressed in glioma cells compared to normal brain, and overexpression of miR-128 is sufficient to repress tumor sphere forming capacity in glioma cells. Another member of the polycomb repressor complexes, EZH2 is under the control of miR-101 in prostate cancer [68]. mIR-101 is lost through genomic deletion in a significant proportion of prostate cancer cells, and its re-expression in cancer cells represses proliferation and migration *in vitro* and tumor growth *in vivo*. Together, these studies suggest that microRNAs may not only be regulating target mRNA translation and stability in cancer stem cells, but also indirectly controlling widespread changes in cellular state through regulation of epigenetic machinery.

The hedgehog (Hh) pathway has a well established role in maintaining stem cell populations [32]. Binding of the soluble Hh ligands to the receptor Patched (Ptc) leads to derepression of the Smoothened (Smo) protein and subsequently to activation of genes regulating proliferation, survival and self renewal, by the Gli transcription factors. The Hh pathway is deregulated in a number of cancers, including medulloblastoma, basal cell carcinoma, breast cancer (our unpublished data) and prostate cancer and is a major target for cancer drug development {Jiang, 2008 #5278}. It is also implicated in maintenance of the CSC phenotype [41]. Recent work has identified microRNA-dependent regulation of several Hh pathways components. In particular, mIR-125b and miR-326 repress expression of Smo, while mir-324–5p targets the Gli1 transcription factor, thus suppressing pathway activation [15]. These microRNAs are downregulated in medulloblastoma, permitting tumor cell proliferation. Similarly, Garzia and colleaugues report that miR-199–5p is under-expressed in CD133-positive medulloblastoma CSCs, and restoration of its expression reduced self renewal and tumor-initiating capacity [17]. Amongst it's targets is HES1, an effector of Notch signalling, which is required for HH-dependent medulloblastomagenesis [20]. There is a developing literature implicating Notch signalling in the CSC phenotype [21].

The miR-34 family of microRNAs are tumor suppressor genes and key transcriptional targets of p53 in the control of proliferation and apoptotic cell death [5, 55]. Recent data suggests a role for miR-34 in the suppression of self renewal and tumor initiating capacity. miR-34 is underexpressed in FACS-purified CSCs from pancreatic [31] and bladder [30] cancer cell lines and its reexpression reduced the proportion of CSCs and the tumor initiating capacity of pancreatic cell lines when xenografted into mice [31]. miR-34 reactivation is associated with dramatic downregulation of Notch proteins and Bcl-2, however the relative importance of these targets to mir-34 action is yet to be elucidated. P53 can regulate the acquisition of self renewal capacity in both iPs cells [24] and mammary stem cells [9], and it remains to be seen whether miR-34 acts downstream of p53 in these contexts.

In hepatocellular carcinoma (HCC), members of the miR-181 family are highly expressed in cancers with a CSC phenotype, characterised by positivity for Epcam and Alpha feta protein (AFP) [29]. Inhibition of mIR-181 reduced the self renewal and tumor initiating capacity of HCC cells and was associated with increases in transcription factors controlling hepatic differentiation, including CDX2 and GATA6, as well as Nemo-like kinase, a suppressor of Wnt signalling. Interestingly, miR-181 expression correlated with the abundance of primitive stem and progenitor cells during normal liver development, suggesting that miR-181 may be another example

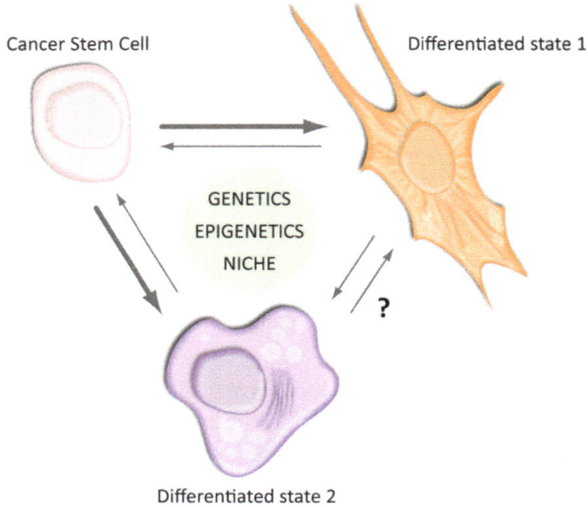

Fig. 1 Schematic representation of the forces driving the maintenance of the CSC phenotype. Self renewal and differentiation of CSCs to progeny may be driven by both cell-autonomous genetic and epigenetic programs as well as by interactions with the local micro-environment, or niche. microRNAs play a key role in this process and are themselves controlled by genetic, epigenetic and environmental factors. Cells with a cancer stem cell phenotype can most likely also be generated by 'de-differentiation', just as iPS cells can be generated from differentiated cells given the appropriate cues. How effective pathophysiological conditions in a tumor are at driving these processes is currently debated

of a microRNAs whose expression in cancer cells with stem-like characteristics is a remnant of its normal physiological function.

4 The cell of origin for Cancer Stem Cells

In many such cases, conservation of CSC-associated microRNAs and their targets between normal and cancer stem cells suggests that the mechanisms regulating both cell types overlap significantly. Furthermore, it raises the possibility that the microRNAome of the CSC is either inherted from an undifferentiated CSC precursor, or that cancer cells 'de-differentiate' to form a CSC population expressing genes and microRNAs in a way that resembles normal stem and progenitor cells. These are important questions for the field to resolve in coming years.

There is little evidence available to discern whether CSC have arisen through the acquisition of mutations in stem or progenitor cells. It is also possible that more differentiated cells have acquired the CSC phenotype through mutation or through interactions with appropriate environmental cues in the tumor microenvironment (Fig. 1). The development of induced pluripotent stem (iPS) cell technology has demonstrated the incredible phenotypic plasticity that resides in all cells. Expression of only 3–4 genes is sufficient to reprogram terminally differentiatied cells, such as fibroblasts, into pluripotent stem cells capable of generating all the tissues of a mature mouse [14, 63]. A number of ES-cell specific microRNAs (miR-291, miR-294, MiR-295), when overexpressed cooperate with Klf4, Oct4 and Sox2 in generating induced pluripotent stem cells (iPS). Interestingly, these microRNAs share seed similarities with a number of microRNAs found deregulated in cancer, (such as the miR-17-92 cluster), thus deregulation of key microRNAs may be sufficient to drive reprogram differentiated cells to a CSC phenotype.

Clearly then, differentiated cells within a tumor may be able to transition to a less differentiated, self renewing population under appropriate conditions. If cells can transition into the CSC phenotype, this may have significant clinical repercussions, as it suggests that therapeutic targeting of CSCs will not be sufficient for long term remission, as CSC may 're-form' from residual cells. However, experimental data to address this issue is limited and at times contradictory. While some studies find that non-CSCs are unable to transition to a CSC phenotype [3, 6, 61], others find that cells expressing CSC markers can form from cells lacking expression of CSC markers [46, 56]. Whether this represents true acquisition of the CSC phenotype, or just the expression of certain cell surface markers, is unclear.

5 microRNAs in Normal Stem Cells and Development

Recent advances in our understanding of the role for microRNAs in embryonic stem (ES) cell biology have added greatly to our understanding of the role of microRNAs in lineage commitment and differentiation. These findings provide a lens with which to view the role of microRNAs in cancer and CSCs. DGCR8, along with Drosha, is a component of the microprocessor complex and required for processing of microRNA primary transcripts to precursors [71], while dicer is required for processing of precursors into mature microRNAs [48]. Surprising recent data demonstrates that microRNA biogenesis is globally suppressed in oocytes and very early embryos [64]. Furthermore, by analysing mice deficient for DGCR8, Suh and colleagues show that microRNAs are not required for early pre-implantation development [64].

ES cells, which are isolated from the inner cell mass of the blastocyst following implantation, express a relatively limited repertoire of microRNAs [25]. We know that these microRNAs are required for ES cell homeostasis, because mouse ES cells carrying homozygous deletion of either DGCR8 [71] or dicer [33] have a defect in both proliferation and differentiation. Furthermore, *In vivo*, deletion of dicer or DGCR8 are embryonic lethal and characterised by failed differentiation.

Some of the microRNAs expressed by ES cells regulate proliferation [70] and self renewal [44], the so-called ES-specific cell cycle (ESCC) microRNAs. Let-7 is also lowly expressed by embryonic stem cells, and acts to repress pluripotency by targeting multiple genes including c-Myc [44]. Interestingly, Let-7 and the ESCC microRNAs act antagonistically, so that expression of the ESCC microRNAs prevents maturation of the Let-7 microRNAs [44]. This process can be uncoupled by studying their function in DGCR8 null cells in which the expression of endogenous microRNAs is absent. This concept of opposing microRNA functions may be an important theme in cancer biology too, and the use of cells in which endogenous microRNAs have been ablated is a powerful approach to study these phenomena.

We have recently shown that miR-380-5p is required for mouse ES cell homeostasis and provides a constitutive survival function by repressing p53 expression [65]. This is, to the best of our knowledge, the first report of a single microRNA required for stem cell viability. miR-380-5p is encoded from a large miRNA cluster found in an imprinted region of human 14q32 [57]. Recently this locus has also been shown to be important for the reprogramming of mouse fibroblasts into induced pluripotent stem (iPS) cells that are competent to give rise to a whole mouse [62]. The expression of transcripts from this region (as detected in ES cells) distinguishes iPS cells that will successfully contribute to chimeric mice from genetically identical iPS cells [62]. Knockdown of p53 is reported to assist with reprogramming [34] so we speculate that expression of miRNAs, such as miR-380-5p, may allow temporary and tunable repression of p53 in stem cells, thus permitting rapid cellular proliferation and self renewal, without the risks associated with irreversible loss of p53 function. We also identified a role for miR-380-5p in cancer, and its expression is required for neuroblastoma survival and proliferation *in vitro* and *in vivo* [65]. Whether miR-380-5p expression plays a role in the iPS process or the cancer stem cell phenotype remains to be elucidated.

Overall, these data suggest that microRNAs are not required for, nor expressed by, the most primitive stem cell but rather are required for lineage specification, consistent with the distinct microRNA expression profiles of different normal cellular lineages [45]. Thus microRNAs may play an important role in cell fate decisions, perhaps by coordinating the expression of suites of genes required to drive these processes, and the net effect of microRNA deficiency is a failure to differentiate. This idea has been confirmed by numerous studies (e.g. [8, 28]).

The direct relevance of these findings to cancer biology and CSC function is yet to be fully elucidated. However, it is interesting to note that, like the diverse microRNA expression between normal tissues, microRNAs are also expressed in very tumor-type specific patterns [42] and that microRNA biogenesis is globally reduced in most cancers [42]. This is most likely contributing to malignancy, since loss of either Dicer [37] or Drosha [36] can promote transformation and tumorigenesis. Could cancers acquire stem-like characteristics by losing the expression of microRNAs that promote lineage commitment and differentiation? In support of this concept, it is interesting to note that of the many microRNAs implicated in CSC biology and described above, the large majority are downregulated in CSCs, while relatively few are gained.

6 Regulation of CSC-Associated microRNAs

The mechanisms by which microRNAs are regulated in CSCs are poorly understood, however in certain cases, post-transcriptional mechanisms may determine the impact of microRNA function on CSC phenotype. Processing and maturation of several microRNAs, including the Let-7 family is controlled by the Lin28 RNA binding protein, which binds to microRNA precursors and inhibits their maturation [22, 23, 69]. Evidence is emerging that pathways controlling EMT and CSC phenotypes, such as receptor tyrosine kinase (RTK) signalling, can impact on microRNA expression. Src family kinases act downstream of extracellular signalling cues such as RTK activation and integrin engagement. Src activation triggers NfKb transcription which can upregulate Lin28 expression [27]. The consequence is a reduction in Let7 maturation and increases in key targets such as high mobility group A2 (HMG2A), Ras and Interleukin-6 (IL-6). These gene expression changes subsequently drive transformation and increases in CSC-associated phenotypes such as tumorsphere forming capacity. Myc is also known to control a number of microRNAs with roles in CSC biology. Myc represses expression of the tumor suppressive microRNAs miR-34a and Let-7 [4], while promoting expression of the members of the miR-17-92 cluster [49], which have been implicated in leukemic CSC function [73]. Myc itself is highly sensitive to environmental signals such as cytokines, growth factors and cell-cell contact [11]. These findings provide a model in which micro-environmental cues, such as RTK activation or the production of inflammatory cytokines are integrated with transcriptional and epigenetic programs and oncogenic signalling to control CSC phenotype via changes in microRNA activity (Fig. 1).

7 Conclusions

There is a large body of literature documenting the aberrant expression and activation of developmental stem cell pathways, such as Wnt, Notch and Hh in cancer and CSCs [66]. Recent data implicates microRNAs as key controllers of the stem cell phenotype and as central players in CSC biology. Many question remain for the field, including how are microRNAs regulated? What are their key targets? And perhaps most importantly, how can the microRNAs critical for CSC function be targeted in a therapeutic context?

Acknowledgments Alex Swarbrick is the recipient of an Early Career Fellowship from the National Breast Cancer Foundation Australia.

References

1. Allan AL, Vantyghem SA, Tuck AB, Chambers AF (2006) Tumor dormancy and cancer stem cells: implications for the biology and treatment of breast cancer metastasis. Breast Dis 26:87–98
2. Boiko AD, Razorenova OV, van de Rijn M, Swetter SM, Johnson DL, Ly DP, Butler PD, Yang GP, Joshua B, Kaplan MJ, et al (2010) Human melanoma-initiating cells express neural crest nerve growth factor receptor CD271. Nature 466:133–137

3. Bonnet D, Dick JE (1997) Human acute myeloid leukemia is organized as a hierarchy that originates from a primitive hematopoietic cell. Nat Med 3:730–737
4. Chang T-C, Yu D, Lee Y-S, Wentzel EA, Arking DE, West KM, Dang CV, Thomas-Tikhonenko A, Mendell JT (2008) Widespread microRNA repression by Myc contributes to tumorigenesis. Nat Genet 40:43–50
5. Chang TC, Wentzel EA, Kent OA, Ramachandran K, Mullendore M, Lee KH, Feldmann G, Yamakuchi M, Ferlito M, Lowenstein CJ, et al (2007) Transactivation of miR-34a by p53 Broadly Influences Gene Expression and Promotes Apoptosis. Mol Cell 26:745–752
6. Charafe-Jauffret E, Ginestier C, Birnbaum D (2009) Breast cancer stem cells: tools and models to rely on. BMC Cancer 9:202
7. Chen CR, Kang Y, Massague J (2001) Defective repression of c-myc in breast cancer cells: A loss at the core of the transforming growth factor beta growth arrest program. Proceedings of the National Academy of Sciences of the United States of America 98, pp 992–999
8. Chen CZ, Li L, Lodish HF, Bartel DP (2004) MicroRNAs modulate hematopoietic lineage differentiation. Science 303:83–86
9. Cicalese A, Bonizzi G, Pasi CE, Faretta M, Ronzoni S, Giulini B, Brisken C, Minucci S, Di Fiore PP, Pelicci PG (2009) The Tumor Suppressor p53 Regulates Polarity of Self-Renewing Divisions in Mammary Stem Cells. Cell 138:1083–1095
10. Dalerba P, Clarke MF (2007) Cancer stem cells and tumor metastasis: first steps into uncharted territory. Cell Stem Cell 1:241–242
11. Dean M, Levine RA, Ran W, Kindy MS, Sonenshein GE, Campisi J (1986) Regulation of c-myc transcription and mRNA abundance by serum growth factors and cell contact. J Biol Chem 261:9161–9166
12. Dykxhoorn DM, Wu Y, Xie H, Yu F, Lal A, Petrocca F, Martinvalet D, Song E, Lim B, Lieberman J (2009) miR-200 enhances mouse breast cancer cell colonization to form distant metastases. PLoS ONE 4:e7181
13. Elenbaas B, Spirio L, Koerner F, Fleming MD, Zimonjic DB, Donaher JL, Popescu NC, Hahn WC, Weinberg RA (2001) Human breast cancer cells generated by oncogenic transformation of primary mammary epithelial cells. Genes Dev 15:50–65
14. Eminli S, Foudi A, Stadtfeld M, Maherali N, Ahfeldt T, Mostoslavsky G, Hock H, Hochedlinger K (2009) Differentiation stage determines potential of hematopoietic cells for reprogramming into induced pluripotent stem cells. Nat Genet 41:968–976
15. Ferretti E, De Smaele E, Miele E, Laneve P, Po A, Pelloni M, Paganelli A, Di Marcotullio L, Caffarelli E, Screpanti I, et al (2008) Concerted microRNA control of Hedgehog signalling in cerebellar neuronal progenitor and tumour cells. EMBO J 27:2616–2627
16. Fusco A, Fedele M (2007) Roles of HMGA proteins in cancer. Nat Rev Cancer 7:899–910
17. Garzia L, Andolfo I, Cusanelli E, Marino N, Petrosino G, De Martino D, Esposito V, Galeone A, Navas L, Esposito S, et al (2009) MicroRNA-199b-5p impairs cancer stem cells through negative regulation of HES1 in medulloblastoma. PLoS ONE 4:e4998
18. Godlewski J, Nowicki MO, Bronisz A, Williams S, Otsuki A, Nuovo G, Raychaudhury A, Newton HB, Chiocca EA, Lawler S (2008) Targeting of the Bmi-1 oncogene/stem cell renewal factor by microRNA-128 inhibits glioma proliferation and self-renewal. Cancer Res 68:9125–9130
19. Gregory PA, Bert AG, Paterson EL, Barry SC, Tsykin A, Farshid G, Vadas MA, Khew-Goodall Y, Goodall GJ (2008) The miR-200 family and miR-205 regulate epithelial to mesenchymal transition by targeting ZEB1 and SIP1. Nat Cell Biol 10:593–601
20. Hallahan AR, Pritchard JI, Hansen S, Benson M, Stoeck J, Hatton BA, Russell TL, Ellenbogen RG, Bernstein ID, Beachy PA, et al (2004) The SmoA1 mouse model reveals that notch signaling is critical for the growth and survival of sonic hedgehog-induced medulloblastomas. Cancer Res 64:7794–7800
21. Harrison H, Farnie G, Brennan KR, Clarke RB (2010) Breast cancer stem cells: something out of notching? Cancer Res 70:8973–8976
22. Heo I, Joo C, Cho J, Ha M, Han J, Kim VN (2008) Lin28 mediates the terminal uridylation of let-7 precursor MicroRNA. Mol Cell 32:276–284

23. Heo I, Joo C, Kim Y-K, Ha M, Yoon M-J, Cho J, Yeom K-H, Han J, Kim VN (2009) TUT4 in Concert with Lin28 Suppresses MicroRNA Biogenesis through Pre-MicroRNA Uridylation. Cell 138:696–708
24. Hong H, Takahashi K, Ichisaka T, Aoi T, Kanagawa O, Nakagawa M, Okita K, Yamanaka S (2009) Suppression of induced pluripotent stem cell generation by the p53-p21 pathway. Nature 460:1132–1135
25. Houbaviy HB, Murray MF, Sharp PA (2003) Embryonic stem cell-specific MicroRNAs. Dev Cell 5:351–358
26. Ibarra I, Erlich Y, Muthuswamy SK, Sachidanandam R, Hannon GJ (2007) A role for microRNAs in maintenance of mouse mammary epithelial progenitor cells. Genes Dev 21:3238–3243
27. Iliopoulos D, Hirsch HA, Struhl K (2009) An epigenetic switch involving NF-kappaB, Lin28, Let-7 MicroRNA, and IL6 links inflammation to cell transformation. Cell 139:693–706
28. Ivey KN, Muth A, Arnold J, King FW, Yeh RF, Fish JE, Hsiao EC, Schwartz RJ, Conklin BR, Bernstein HS, et al (2008) MicroRNA regulation of cell lineages in mouse and human embryonic stem cells. Cell Stem Cell 2:219–229
29. Ji J, Yamashita T, Budhu A, Forgues M, Jia HL, Li C, Deng C, Wauthier E, Reid LM, Ye QH, et al (2009a) Identification of microRNA-181 by genome-wide screening as a critical player in EpCAM-positive hepatic cancer stem cells. Hepatology 50:472–480
30. Ji Q, Hao X, Meng Y, Zhang M, Desano J, Fan D, Xu L (2008) Restoration of tumor suppressor miR-34 inhibits human p53-mutant gastric cancer tumorspheres. BMC Cancer 8:266
31. Ji Q, Hao X, Zhang M, Tang W, Yang M, Li L, Xiang D, Desano JT, Bommer GT, Fan D, et al (2009b) MicroRNA miR-34 inhibits human pancreatic cancer tumor-initiating cells. PLoS ONE 4:e6816
32. Jiang J, Hui CC (2008) Hedgehog signaling in development and cancer. Dev Cell 15:801–812
33. Kanellopoulou C, Muljo SA, Kung AL, Ganesan S, Drapkin R, Jenuwein T, Livingston DM, Rajewsky K (2005) Dicer-deficient mouse embryonic stem cells are defective in differentiation and centromeric silencing. Genes Dev 19:489–501
34. Kawamura T, Suzuki J, Wang YV, Menendez S, Morera LB, Raya A, Wahl GM, Belmonte JC (2009) Linking the p53 tumour suppressor pathway to somatic cell reprogramming. Nature 460:1140–1144
35. Korpal M, Lee ES, Hu G, Kang Y (2008) The miR-200 family inhibits epithelial-mesenchymal transition and cancer cell migration by direct targeting of E-cadherin transcriptional repressors ZEB1 and ZEB2. J Biol Chem 283:14910–14914
36. Kumar MS, Lu J, Mercer KL, Golub TR, Jacks T (2007) Impaired microRNA processing enhances cellular transformation and tumorigenesis. Nat Genet 39:673–677
37. Kumar MS, Pester RE, Chen CY, Lane K, Chin C, Lu J, Kirsch DG, Golub TR, Jacks T (2009) Dicer1 functions as a haploinsufficient tumor suppressor. Genes Dev 23:2700–2704
38. Lapidot T, Sirard C, Vormoor J, Murdoch B, Hoang T, Caceres-Cortes J, Minden M, Paterson B, Caligiuri MA, Dick JE (1994) A cell initiating human acute myeloid leukaemia after transplantation into SCID mice. Nature 367:645–648
39. Li Y, VandenBoom TG, 2nd Kong D, Wang Z, Ali S, Philip PA, Sarkar FH (2009) Up-regulation of miR-200 and let-7 by natural agents leads to the reversal of epithelial-to-mesenchymal transition in gemcitabine-resistant pancreatic cancer cells. Cancer Res 69:6704–6712
40. Lindeman GJ, Visvader JE (2010) Insights into the cell of origin in breast cancer and breast cancer stem cells. Asia Pac J Clin Oncol 6:89–97
41. Liu S, Dontu G, Mantle ID, Patel S, Ahn N-s, Jackson KW, Suri P, Wicha MS (2006) Hedgehog Signaling and Bmi-1 Regulate Self-renewal of Normal and Malignant Human Mammary Stem Cells. Cancer Res 66:6063–6071
42. Lu J, Getz G, Miska EA, Alvarez-Saavedra E, Lamb J, Peck D, Sweet-Cordero A, Ebert BL, Mak RH, Ferrando AA, et al (2005) MicroRNA expression profiles classify human cancers. Nature 435:834–838

43. Mani SA, Guo W, Liao MJ, Eaton EN, Ayyanan A, Zhou AY, Brooks M, Reinhard F, Zhang CC, Shipitsin M, et al (2008) The epithelial-mesenchymal transition generates cells with properties of stem cells. Cell 133:704–715
44. Melton C, Judson RL, Blelloch R (2010) Opposing microRNA families regulate self-renewal in mouse embryonic stem cells. Nature 463:621–626
45. Mendell JT (2005) MicroRNAs: critical regulators of development, cellular physiology and malignancy. Cell Cycle 4:1179–1184
46. Meyer M, Fleming J, Ali M, Pesesky M, Ginsburg E, Vonderhaar B (2009) Dynamic regulation of CD24 and the invasive, CD44posCD24neg phenotype in breast cancer cell lines. Breast Cancer Res 11:R82
47. Morel AP, Lievre M, Thomas C, Hinkal G, Ansieau S, Puisieux A (2008) Generation of breast cancer stem cells through epithelial-mesenchymal transition. PLoS ONE 3:e2888
48. Murchison EP, Partridge JF, Tam OH, Cheloufi S, Hannon GJ (2005). Characterization of Dicer-deficient murine embryonic stem cells. PNAS 102:12135–12140
49. O'Donnell KA, Wentzel EA, Zeller KI, Dang CV, Mendell JT (2005) c-Myc-regulated microRNAs modulate E2F1 expression. Nature 435:839–843
50. Park IK, Morrison SJ, Clarke MF (2004) Bmi1, stem cells, and senescence regulation. J Clin Invest 113:175–179
51. Park SM, Gaur AB, Lengyel E, Peter ME (2008) The miR-200 family determines the epithelial phenotype of cancer cells by targeting the E-cadherin repressors ZEB1 and ZEB2. Genes Dev 22:894–907
52. Polyak K. Haviv I, Campbell IG (2009) Co-evolution of tumor cells and their microenvironment. Trends in Genetics 25:30–38
53. Quintana E, Shackleton M, Sabel MS, Fullen DR, Johnson TM, Morrison SJ (2008) Efficient tumour formation by single human melanoma cells. Nature 456:593–598
54. Rajasekhar VK, Begemann M (2007) Concise review: roles of polycomb group proteins in development and disease: a stem cell perspective. Stem Cells 25:2498–2510
55. Raver-Shapira N, Marciano E, Meiri E, Spector Y, Rosenfeld N, Moskovits N, Bentwich Z, Oren M (2007) Transcriptional Activation of miR-34a Contributes to p53-Mediated Apoptosis. Mol Cell 26:731–743
56. Roesch A, Fukunaga-Kalabis M, Schmidt EC, Zabierowski SE, Brafford PA, Vultur A, Basu D, Gimotty P, Vogt T, Herlyn M (2010) A temporarily distinct subpopulation of slow-cycling melanoma cells is required for continuous tumor growth. Cell 141:583–594
57. Seitz H, Royo H, Bortolin M-L, Lin S-P, Ferguson-Smith AC, Cavaille J (2004) A Large Imprinted microRNA Gene Cluster at the Mouse Dlk1-Gtl2 Domain. Genome Res 14:1741–1748
58. Shimono Y, Zabala M, Cho RW, Lobo N, Dalerba P, Qian D, Diehn M, Liu H, Panula SP, Chiao E, et al (2009) Downregulation of miRNA-200c links breast cancer stem cells with normal stem cells. Cell 138:592–603
59. Shipitsin M, Campbell LL, Argani P, Weremowicz S, Bloushtain-Qimron N, Yao J, Nikolskaya T, Serebryiskaya T, Beroukhim R, Hu M, et al (2007) Molecular definition of breast tumor heterogeneity. Cancer Cell 11:259–273
60. Singh A, Settleman J (2010) EMT, cancer stem cells and drug resistance: an emerging axis of evil in the war on cancer. Oncogene 29:4741–4751
61. Somervaille TC, Matheny CJ, Spencer GJ, Iwasaki M, Rinn JL, Witten DM, Chang HY, Shurtleff SA, Downing JR, Cleary ML (2009) Hierarchical maintenance of MLL myeloid leukemia stem cells employs a transcriptional program shared with embryonic rather than adult stem cells. Cell Stem Cell 4:129–140
62. Stadtfeld M, Apostolou E, Akutsu H, Fukuda A, Follett P, Natesan S, Kono T, Shioda T, Hochedlinger K (2010) Aberrant silencing of imprinted genes on chromosome 12qF1 in mouse induced pluripotent stem cells. Nature 465:175–181
63. Stadtfeld M, Nagaya M, Utikal J, Weir G, Hochedlinger K (2008) Induced pluripotent stem cells generated without viral integration. Science 322:945–949

64. Suh N, Baehner L, Moltzahn F, Melton C, Shenoy A, Chen J, Blelloch R (2010) MicroRNA Function Is Globally Suppressed in Mouse Oocytes and Early Embryos. Current Biol 20:271–277
65. Swarbrick A, Woods SL, Shaw A, Balakrishnan A, Phua Y, Nguyen A, Chanthery Y, Lim L, Ashton LJ, Judson RL, et al (2010). miR-380-5p represses p53 to control cellular survival and is associated with poor outcome in MYCN-amplified neuroblastoma. Nat Med 16:1134–1140
66. Taipale J, Beachy PA (2001) The Hedgehog and Wnt signalling pathways in cancer. Nature 411:349–354
67. Thiery JP, Acloque H, Huang RY, Nieto MA (2009) Epithelial-mesenchymal transitions in development and disease. Cell 139:871–890
68. Varambally S, Cao Q, Mani RS, Shankar S, Wang X, Ateeq B, Laxman B, Cao X, Jing X, Ramnarayanan K, et al (2008) Genomic loss of microRNA-101 leads to overexpression of histone methyltransferase EZH2 in cancer. Science 322:1695–1699
69. Viswanathan SR, Daley GQ, Gregory RI (2008) Selective blockade of microRNA processing by Lin28. Science 320:97–100
70. Wang Y, Baskerville S, Shenoy A, Babiarz JE, Baehner L, Blelloch R (2008) Embryonic stem cell-specific microRNAs regulate the G1-S transition and promote rapid proliferation. Nat Genet 40:1478–1483
71. Wang Y, Medvid R, Melton C, Jaenisch R, Blelloch R (2007) DGCR8 is essential for microRNA biogenesis and silencing of embryonic stem cell self-renewal. Nat Genet 39:380–385
72. Wellner U, Schubert J, Burk UC, Schmalhofer O, Zhu F, Sonntag A, Waldvogel B, Vannier C, Darling D, zur Hausen A, et al (2009) The EMT-activator ZEB1 promotes tumorigenicity by repressing stemness-inhibiting microRNAs. Nat Cell Biol 11:1487–1495
73. Wong P, Iwasaki M, Somervaille TC, Ficara F, Carico C, Arnold C, Chen CZ, Cleary ML (2010) The miR-17-92 microRNA polycistron regulates MLL leukemia stem cell potential by modulating p21 expression. Cancer Res 70:3833–3842
74. Yin G, Chen R, Alvero AB, Fu HH, Holmberg J, Glackin C, Rutherford T, Mor G (2010) TWISTing stemness, inflammation and proliferation of epithelial ovarian cancer cells through MIR199A2/214. Oncogene 29:3545–3553
75. Yu F, Deng H, Yao H, Liu Q, Su F, Song E (2010) Mir-30 reduction maintains self-renewal and inhibits apoptosis in breast tumor-initiating cells. Oncogene 29:4194–4204
76. Yu F, Yao H, Zhu P, Zhang X, Pan Q, Gong C, Huang Y, Hu X, Su F, Lieberman J, et al (2007) let-7 regulates self renewal and tumorigenicity of breast cancer cells. Cell 131:1109–1123

MicroRNAs in the Pathogenesis of Viral Infections and Cancer

Derek M. Dykxhoorn

Abstract MicroRNAs (miRNAs) have emerged as a class of broadly conserved small RNAs that facilitate the sequence-specific, post-transcriptionally regulation of gene expression. These small regulatory RNAs have been shown to play essential roles in many important biological processes from metabolism to apoptosis. Alterations in miRNA expression profiles can have pathogenic consequences. This article will examine the role that miRNAs play in the pathophysiology of viral infections and cancer.

1 Introduction

MicroRNAs (miRNAs) are endogenous ~22-nucleotide (nt) RNAs that regulate gene expression by binding to the specific sites on the mRNA of protein coding genes to direct their repression. [1–3] The first hints that small RNA molecules could regulate gene expression came with the discovery of the first miRNAs, lin-4 and let-7, which were identified genetically based on their role in regulating developmental timing in *Caenorhabditis elegans*. [4–6] Homologs of let-7 were soon identified in a diverse array of organisms including *Drosophila melanogaster* (fruitfly), *Danio rerio* (Zebrafish), and humans [7]. These miRNAs have been found to be representatives of a much larger class of small regulatory RNAs that are conserved from algae to humans [8, 9]. miRNA genes are distributed throughout the human genome with the exception of the Y chromosome [10]. MiRNAs are often found in clusters where multiple miRNAs, often related in sequence, are expressed in the same transcriptional units. Although the majority of miRNA were thought to be encoded in intergenic regions, recent studies have shown that many are located in protein coding genes (either in introns or less commonly in exons) [11]. In fact, the

D. M. Dykxhoorn (✉)
Dr. John T. Macdonald Foundation of Human Genetics and the Department of Microbiology and Immunology, John P. Hussman Institute for Human Genomics,
University of Miami Miller School of Medicine, Miami, FL 33136, USA
e-mail: DDykxhoorn@med.miami.edu

S. Alahari (ed.), *MicroRNA in Cancer*,
DOI 10.1007/978-94-007-4655-8_4,

majority of mammalian miRNAs are present within defined transcriptional units. With few exceptions, miRNAs are transcribed by RNA polymerase II producing a transcript containing a 7-methylguanosine cap and a poly (A) tail [12, 13]. This permits the developmental and tissue-specific expression of miRNAs, for example, let-7 is expressed during the L3 stage in C. elegans and facilitates the switch from embryonic to adult developmental stages, [6, 7] while the expression of miR-1 is restricted to muscle cells [14].

Transcription of miRNA genes yields long primary transcripts (pri-miRNAs) which contain one to several local fold back structures (Fig. 1) [15, 16]. The mature ~22-nt miRNA is derived from these stem loop structures within the pri-miRNA through the step wise processing by members of the RNase III-family of endoribonucleases, Drosha and Dicer. Rather than recognize specific sequence motifs, Drosha, in concert with the double stranded RNA (dsRNA) binding protein DGCR8 (DiGeorge syndrome critical region gene 8), recognizes structural elements of the stem loop [17–22]. Specifically, Drosha and DGCR8 (collectively termed the microprocessor complex [17, 23]) recognizes the ~33-bp stem, terminal loop, and a flanking single stranded (ss)RNA segment of the pri-miRNA hairpin. [15, 16, 18, 24] By interacting with the dsRNA stem and the ssRNA flanking sequence, DGCR8 helps to place Drosha in place to cleave the pri-miRNA 11 bp from the ssRNA-dsRNA junction. Accumulated research supports the hypothesis that this initial cleavage event occurs co-transcriptionally. By linking Drosha processing with transcription and splicing facilitates the biogenesis of the intron-encoded miRNA without impacting the levels of the corresponding mRNA (coordinated co-transcriptional process). Drosha cleavage produces an ~70-nt precursor miRNA (pre-miRNA). This precursor is translocated from the nucleus to the cytoplasm by the karyopherin exportin 5 [25–27]. Originally, exportin 5 was believed to act as a minor export factor for tRNAs, however, with the discovery of miRNAs it has become apparent that the major function of exportin 5 is the nuclear export of pre-miRNAs. Once in the cytoplasm, the pre-miRNA is processed by the RNase III-family member Dicer (in concert with the dsRNA binding protein TRBP) to generate the ~22-nt miRNA duplexes containing the mature miRNA [28–30]. One strand of the Dicer product remains as a mature miRNA and is assembled into the effector complex called miRNP or miRNA-induced silencing complex (miRISC) [31, 32]. The thermodynamic properties of the 5′ termini of the duplexed miRNA determines which strand will be incorporated into miRISC [32]. The mature single stranded RISC-associated mature miRNA provides the specificity determinant guiding the effector complex miRISC to the target mRNAs.

Mature miRNAs post-transcriptionally regulate gene expression by binding to complementary sequences found predominantly in the 3′ UTR of target mRNAs [1, 2, 33, 34]. Large scale analysis of protein levels and mRNA levels in cells transiently overexpressing a miRNA or deleted for a specific miRNA revealed that gene silencing resulting from solely translational repression accounted for only a small number of the total targets and was associated with only modest decreases in protein levels [35–39]. Conversely, proteins that show the most robust silencing were a result of transcript destabilization and degradation. This mRNA destabilization

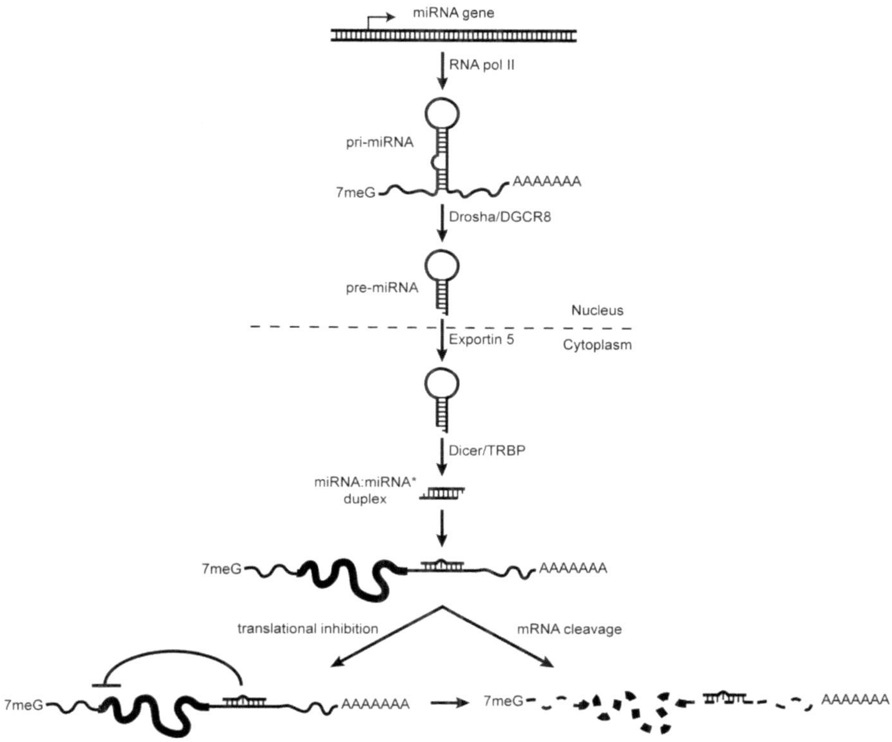

Fig. 1 MicroRNA biogenesis pathway. MicroRNAs direct the posttranscriptional silencing of gene expression are derived from extended primary transcripts predominantly expressed from RNA polymerase II promoters [11, 12, 143]. These primary transcripts, termed pri-miRNAs, encoded one to several highly structured RNA hairpin that contains frequent bulges and mismatches from which mature miRNA(s) are derived [1, 2]. These long hairpins are cleaved into shorter (~70 nt) hairpin RNAs, precursor miRNAs (pre-miRNAs), in the nucleus by Drosha in conjunction with the double-stranded RNA recognition protein, termed DGCR8 in mammalian cells [23, 144–147]. Pre-miRNAs are exported into the cytoplasm by the karyopheron Exportin 5 where they are recognized and cleaved into the ~22 nt microRNA by Dicer in conjunction with another dsRNA-binding protein, TRBP in mammals [25, 26, 29, 148, 149]. The miRNA associates with an argonaute protein which forms the core of the miRNA containing RNA-induced silencing complex (miRISC), and the passenger strand (miRNA*) is lost, leaving the mature microRNA to guide the recognition of the microRNA-binding sites on the target mRNA, leading to silencing of target gene expression [150–152] The binding of the miRISC to the miRNA-binding site on the target mRNAs results in the inhibition of translation and/or mRNA degradation [2, 35, 39]

could result from either the direct cleavage of the target mRNA or the degradation of transcripts that are undergoing translational repression. The principally factor determining the specificity of binding between a miRNA and its target sequence are nucleotides 2–7 of the mature miRNA which constitutes the ‘seed region’ [36, 40–43]. In some cases, binding between the 3′ end of a miRNA with sequences in the miRNA binding site on the target mRNA can compensate for imperfect seed pairing [38, 44].

To date, more than 10,000 miRNAs have been annotated in 96 species, including over 700 human miRNAs (miRbase v. 14.0). The computational prediction of miRNA target sites suggest that greater than 60 % of all human protein-coding genes possess targets of putative (potential) miRNA binding sites [2, 36]. Each miRNA is predicted to regulate the expression of hundreds even thousands of transcripts [45]. This suggests that the dysregulation of expression of even a small number of miRNAs can have profound effects on the pattern of gene expression and thereby the physiology of the cell. Since miRNAs play key regulatory roles in a wide variety of cellular processes, the alteration of miRNA expression patterns can have pathogenic consequences. Rather than serving as a comprehensive compendium of miRNA-mRNA interactions in disease development, this article will focus on the mechanisms by which miRNA expression is dysregulated in viral infections and cancer.

2 Viral miRNAs

RNA-mediated gene silencing has been shown to be an important component in the antiviral defense system of plants and insects [46–50]. This defense system is based on the recognition and endonucleolytic cleavage of long dsRNAs, produced as a consequence of the replication of many viral pathogens that are cleaved into small interfering RNAs (siRNAs) by the RNase III-type enzyme Dicer [51]. These virus-specific siRNAs are incorporated into RISC and direct the cleavage of the complimentary viral messenger RNAs (mRNAs) resulting in the suppression of infection by genetically related viruses. In vertebrates, the immune system has evolved to clear viral and bacterial pathogens. As a consequence, a more complex relationship has evolved between viral pathogens and the RNAi pathway.

MiRNAs have several characteristics that make them an attractive mechanism for viruses to utilize for the regulation of gene expression. Since miRNAs are small in size they can easily be encoded in viral genomes where space is of a premium and they can evolve rapidly to target new transcripts. In addition, miRNAs are non-immunogenic and they take advantage of a ubiquitous, endogenous host regulatory mechanism. Finally, a single miRNA has the potential to alter the expression pattern of a large number of genes. By altering the gene expression patterns in target cells, miRNAs can create a cellular environment conducive to viral replication. There are several ways that miRNAs could be envisioned to function in viral pathogenesis, including the regulation of viral gene expression by virus-encoded miRNAs and the regulation of host genes by virus-encoded miRNAs.

Biochemical and computational approaches have been used to gain a better understanding of the complexity of the small regulatory RNA coding capacity from a wide variety of cell types, organisms, and development stages [52–58]. The majority of virally-encoded miRNAs discovered to date have been identified in herpesviruses, a family of large DNA viruses that establish latent infections in a variety of target cells [59–65]. MiRNA are encoded from the genomes of several herpesviruses that cause disease in humans, including the γ-herpesviruses Epstein Barr virus (EBV) [60, 63, 66] and Kaposi's sarcoma-associated herpes virus (KSHV), [59, 62, 65, 67] which encode 25 and 12 miRNAs respectively. In addition, the human α-herpesviruses herpes simplex virus 1 (HSV-1) [68–70] and Herpes simplex virus-2 (HSV-2) [71–73] have been shown to encode 8 and 6 miRNAs a piece, while the β-herpesvirus human cytomegalovirus (CMV) [74, 75] encodes 11 miRNAs. Computational analysis failed to identify miRNAs from the genomes of several other herpesviruses; the α-herpesvirus HHV-3 (varicella-zoster virus) or the β-herpesvirus HHV-6 and HHV-7. Initially, Pfeffer et al. [59] found that there was a lack of conservation in miRNA sequences in miRNA sequence from different herpesviruses genomes. This suggests that the miRNAs expressed from these different herpesviruses are not involved in core functions of the virus (viral replication, viral gene expression, etc.) but have evolved independently to allow each virus to adapt to the specific cell types in which the virus persists. However, the comparison of 2 more closely related herpesviruses, EBV and Rhesus lymphocryptovirus (rLCV), both members of the lymphocryptovirus genus of herpesviruses, led to the biochemical identification of 7 highly conserved miRNAs between them [63]. These conserved miRNAs were present in both the BHRF1 and BART1 miRNA clusters of the viral genomes (see below) [63, 76]. The high degree of miRNA conservation found between these 2 viruses may be reflective of the similar cellular environments in which these viruses replicate.

Epstein Barr virus (EBV, also called human herpesvirus-4 (HHV-4)), preferentially infects B cells and has been linked with several malignancies, including Hodgkin's lymphoma, Burkitt's lymphoma and nasopharyngeal carcinomas [63, 66, 67]. The EBV-encoded miRNAs were found in three regions within the EBV genome. The largest cluster, containing 21 miRNAs, was mapped to the intronic regions of the Bam HI-A region rightward transcript (BART) gene (miR-BART1 and −3 to −22), while a second site in the intron between BALF5 and LF2 encodes the miR-BART2 [63, 66, 67]. Another cluster, containing the 3 remaining miRNA sequences, was located within the 5′ and 3′ untranslated regions (UTRs) of the BamHI fragment H rightward open reading frame 1 (BHRF1) gene, miR-BHRF1-1 to -3 [60, 63]. Although a role in the maintenance of latency has been suggested for the EBV encoded miRNAs, this has been challenged by the observation that the genome region encoded most of the miR-BARTs (17 miRNAs) is deleted in the laboratory strain B89-5 without compromising the ability of the virus to immortalize B cells [63, 67].

Karposi's sarcoma-associated virus (KHSV), the etiological agent of Kaposi's sarcoma and primary effusion lymphoma, has been shown to encode 12 miRNAs residing within two clusters [59, 65, 67]. The majority of the KSHV-encoded miRNAs (10 miRNAs) are present in the intron between ORF71 and the kaposin gene (K12 gene), while the remaining 2 miRNAs are expressed from the coding region (miR-K10) or 3′UTR of the KHSV K12 gene (miR-K12). A role for the intronic KSHV miRNAs in the maintenance of latency has been suggested by the lack of expression of these miRNAs during lytic infections and the deletion of these miRNAs does not inhibit lytic viral replication [59, 61, 65, 77]. Conversely, both the KSHV miR-K10 and -K12 have been found to be up-regulated in response to lytic induction [62, 67]. The role that these miRNAs play in the pathophysiology of KSHV infection remains largely unknown.

The expression of a non-protein coding RNA in HSV-1 has been associated with the maintenance of latency in trigeminal sensory neurons [78, 79]. This transcript, termed the latency-associated transcript (LAT), is the only viral gene expressed during latency [80]. Computational analysis predicted that a stem-loop structure, reminiscent of the structure of pre-miRNAs, could be formed in exon 1 of the LAT gene bF2 [78, 79, 81]. Next generation sequencing (NGS) analysis of HSV-1 infected mouse and human trigeminal ganglia identified 7 miRNAs encoded from the HSV-1 genome, including six encoded by the LAT [68, 69]. On the other hand, the remaining miRNA (miR-H1) is encoded upstream of LAT and is expressed in productively replicating cells [70]. Cells that express the LAT gene have been shown to be refractory to cisplatin-induced apoptosis but only in the context of a functioning RNAi pathway. The siRNA-mediated silencing of Dicer, which is necessary for the formation of mature miRNAs, inhibited the anti-apoptotic effects of the LAT gene. These effects were also attenuated in HSV-1-infected cells treated with a non-cleavable oligonucleotide complementary to miR-LAT [79]. Five LAT-associated miRNAs have been identified in HSV-2 through NGS of the small RNA pool of HSV-2 infected sacral ganglia [71–73]. An additional miRNA has been identified in HSV-2 encoded from the long unique region (U_L) of the HSV genome [73]. While the majority of miRNAs encoded by the α-herpesviruses (HSV-1 and -2) and γ-herpesviruses (EBV and KSHV) appear to be linked to the maintenance of latency, [59, 63, 65, 68, 69, 71–73, 77] the CMV-encoded miRNAs have been most abundantly found in primary fibroblasts undergoing lytic replication [59, 74, 75]. These miRNAs are spread throughout the CMV genome in small groups, including intergenic and intronic regions. These miRNAs all demonstrate early kinetics, expression during the immediate early or early genes, supporting their role in lytic infections. However, any potential role that these miRNAs are playing in latency may be obscured by a lack of latent CMV infection models.

Murine herpesvirus-68 (MHV-68) represents an unusual class of miRNAs that are expressed from RNA polymerase III (Pol III) and are processed in a Drosha-independent manner [82, 83]. All of the MHV-68 miRNAs (9 in total) are transcribed by RNA pol III from internal transfer RNA (tRNA promoters producing ~130- to 200-nt long pri-miRNAs containing an ~60-nt 5′-tRNA moiety linked to one or two ~70-nt miRNA-encoding hairpins. The cleavage of these pri-miRNAs by tRNase Z releases of the viral pre-miRNAs [83]. These pre-miRNAs join the endogenous miRNA biogenesis pathway leading to their export from the nucleus and cytoplasmic processing by Dicer and the production of at least nine mature miRNAs [82, 83].

One interesting mechanism that has been adopted by several viruses is the expression of miRNA that have homology to cellular miRNAs. This may be an example of viral host co-evolution where the virus has adopted miRNAs with sequence similarity to cellular miRNAs. In this way, the viral miRNAs can regulate a set of target genes that overlaps with their cellular counterparts. For example, the KSHV-encoded miR-K11 contains an identical seed sequence as the cellular miR-155, a miRNA which plays a central role in lymphocyte development and whose dysregulation has been linked to the development of hematopoietic malignancies [84, 85]. Gottwein et al [109] showed that KSHV miR-K11 regulates a common set of genes as miR-155, including the proto-oncogene c-Fos, angiotensin II receptor-associated protein (AGTRAP), and riboflavin kinase (RFK) [77, 86, 87]. In addition to KHSV, the dysregulation of miR-155 expression or expression of miRNAs similar to miR-155 appears to be a common mechanism used by several viruses. Marek's disease virus (MDV) encodes an ortholog of miR-155, [86] while EBV and the chicken retrovirus reticuloendotheliosis virus strain T induce the expression of cellular miR-155 [88, 89]. These examples suggest that alteration of miR-155-regulated genes and pathways may facilitate viral oncogenesis. Several other viruses encode miRNAs that have seed sequences with homology to cellular miRNAs. EBV miR-BART5, MHV-68 miR-M1-7-5p, and rLCV miR-rL1-8 share seed homology with miR-18a/b, [59, 63] while MHV-68 miR-M1-4 shares seed homology with miR-151-5p [90] and EBV-BART1-3p, rLCV miR-rL1-6-5p and MDV-2 miR-M21 share seed homology with miR-29abc [59, 63, 91].

Several other DNA viruses have been shown to encode miRNAs from their genomes. Two miRNAs (termed viral associated RNAs (VA RNA 1 and VA RNA 2)) have been shown to be highly expressed in adenovirus-infected cells [92]. These RNAs are transcribed by RNA pol III and computational analysis of their structure suggested they adopt an imperfect stem-loop secondary structure similar to that found in pre-miRNAs [93]. Interestingly, these highly expressed virally-encoded miRNAs have been suggested to act as inhibitors of the miRNA biogenesis machinery [93]. Similar to pre-miRNAs, they use the nuclear-export receptor exportin 5 for their translocation from the nucleus into the cytoplasm. The VA RNAs are so highly expressed that they effectively compete with the endogenous miRNAs for exportin 5, leading to the saturation of the nuclear translocation activity [93, 94]. Once in the cytoplasm, the VA-encoded pre-miRNAs are recognized and processed by Dicer into mature miRNAs. Although these adenovirus-encoded miRNAs associate with RISC during lytic viral infections, no target genes for these miRNAs have been identified and it is unclear whether they play any physiological function. These highly abundant RNAs appear to competitively bind to Dicer and prevent the processing of endogenous Dicer substrates (i.e. endogenous miRNAs and small interfering (si)RNAs) [93, 94].

The polyoma virus, simian virus 40 (SV40) encodes a single pre-miRNA, miR-S1. This miRNA is expressed in late stages of the infection. The genomic location and expression pattern of miR-S1 are conserved with other polyomaviruses, including Merkel cell virus (MCV), BK virus (BKV), JC virus (JCV), and the primate and mouse polyomaviruses SA12 and mPyV, respectively. It appears that this miRNA, miR-S1, regulates the expression of early viral genes and it has been suggested that it helps to reduce the susceptibility of SV40 infected cells to lysis by cytotoxic T cells [95–98].

To 1° or another, every virus is dependent upon the cellular machinery to complete the viral life cycle. Therefore, complex relationships have evolved between viruses and the host cells they infect to ensure the fitness and survival of the virus. Small regulatory RNAs have added another dimension to these complex viral host interactions. Only a handful of viral miRNa have identified targets. Based on this limited sampling of interactions, it is clear that viral miRNAs function to alter cellular programs that regulate latent-lytic switches, promote cell survival, proliferation and/or differentiation, and modulate the immune response. The identification of viral miRNA targets is complicated by the diversity of targeting interaction that can lead to productive gene silencing. Although the seed sequence of miRNAs (nucleotides 2–8 of the miRNA) plays a key role in target recognition in many cases, it is not universally required by each miRNA-mRNA pairing [2, 99]. For example, miRNA that lack complete seed pairing often contain 3′ compensatory binding interactions. Since miRNAs require such small regions of complementarity to induce translational inhibition and/or mRNA cleavage and degradation, a large number of targets (many of which are not physiological) are predicted for each miRNA [36, 45]. Identification of the relevant target genes from among these large numbers of computationally predicted targets remains a challenge. Despite these challenges, viral miRNAs have been shown to regulate either viral genes or cellular genes.

3 Viral miRNAs Targeting Viral Messages

The identification of viral targets of viral miRNAs is often much easier to do than to identify cellular targets due to the smaller size of the genome and the fact that in many cases the viral miRNAs are expressed from transcripts that overlap with the viral genes that they regulate. The SV40 miRNA, miR-S1 is expressed antisense from the viral T-antigen (T-Ag) mRNA [97]. The expression of miR-S1 suppresses T-Ag expression late in viral infection. The inhibition of miR-S1 expression results in elevated levels of T-Ag and increased killing by CD8+cytotoxic T cells. The common location and expression pattern of corresponding miRNAs from other polyoma viruses suggest that this downregulation of T-Ag expression is conserved across the various polyomavirus families and serves a protective function to inhibit T-Ag-dependent activation of cytotoxic T cells [95, 96, 98, 100]. In another example, the EBV miR-BART2 is encoded antisense to the DNA polymerase BALF5 gene [101]. During latent infection, the miR-BART2 is highly expressed and silences BALF5 expression, while the induction of lytic replication leads to a decrease in miR-BART2 derepressing BALF5 expression. Therefore, it appears that miR-BART2 regulates the switch between the latent and lytic phases of the EBV lifecycle. Several miRNAs from HSV-1 and -2 lie antisense to viral protein encoding transcripts [68, 71–73]. The expression of both HSV-1 and -2 miR-H2 are expressed antisense to the immediate early viral gene ICP0 whose expression is down-regulated by these miRNAs. Similarly, the HSV-1 and -2 miRNAs, miR-H3 and H4 are expressed from transcripts that are overlapping and antisense to the viral ICP34.5 gene. The over-expression of HSV-2 miR-H3 led to the down-regulation of ICP34.5 expression. ICP34.5 serves as a neurovirulence factor that appears to protect herpesvirus infected neurons from the induction of apoptosis by inhibiting the activity of the double stranded RNA-activated PKR (protein kinase R) [102].

In addition to targeting overlapping viral protein encoding transcripts, several viral miRNAs target viral genes containing complementary binding sites with which they are discontinuous. The HSV-1 miRNA, miR-H6 can bind to complementary sites on the viral ICP4 mRNA inhibiting ICP4 protein expression [68]. ICP4 is a transcriptional transactivator that plays an important role in the induction of lytic viral replication. Similarly, two miRNAs encoded in the chicken α-herpesvirus infectious laryngotracheitis virus (ILTV) also appear to silence ICP4 expression suggesting this is a general mechanism used by herpes viruses for the regulation of lytic infections [91]. Similarly, the hCMV miR-UL112-1 can bind to two sites on the 3′ UTR of the viral IE72/IE1 gene [75, 103–105]. Transgenic over-expression of miR-UL112-1 in cells prior to CMV infection reduced viral replication levels [104, 105]. KSHV miR-K9* targets the silencing of the Replication Transactivation Activator (RTA or ORF50) which plays a central role in regulating the switch between latent and lytic viral replication by activating a cascade of viral protein expression leading to the production of the components of the viral capsid and the replication of the viral genome [106]. By suppressing RTA expression, miR-K9* regulates the entry into the lytic cycle.

4 Viral miRNAs that Regulate Host Genes

A second mechanism that can be used by viruses to alter target cell physiology to promote viral infection and replication is the regulation of cellular genes by virally-encoded miRNAs. Potential cellular targets for viral miRNAs could be cellular genes involved in antiviral defense systems, cell proliferation and survival, immune recognition and stress responses. The inhibition of apoptosis is a common mechanism used by viruses to ensure their survival. The expression of the non-coding RNA, latency-associated transcript (LAT), from the HSV-1 genome has been associated with the maintenance of viral latency in neuronal cells. Biochemical experiments have demonstrated that this transcript produces a mature miRNA. Cells that express LAT are refractory to cisplatin-induced apoptosis [79]. The targeted inhibition of LAT using non-cleavable oligonucleotides complementary to the mature LAT encoded miRNA attenuated the anti-apoptotic effects of LAT. Computational approaches identified potential LAT miRNA binding sites within the 3′ UTR of two apoptosis-associated genes, transforming growth factor-β (TGF-β) and mother against decapentaplegic homolog 3 (SMAD3) [79]. Consistently, expression of both of these transcripts is down-regulated in cells infected with wild-type HSV-1 but not in cells infected with a mutant HSV-1 containing a deletion in the region from which the LAT miRNA is encoded.

One of the most well characterized viral miRNA-host target gene interactions is between the KHSV miRNAs and Thrombospondin 1 (THBS1), a multidomain matrix glycoprotein that mediates cell-to-cell and cell-to-matrix interactions [107]. The overexpression of multiple KSHV pre-miRNAs (pre-miR-K1 to K9 and K11) resulted in reduced THBS1 expression at the mRNA and protein levels. Since

THBS1 acts to inhibit angiogenesis and cell growth by interacting with TGF-β, the KHSV miRNA-mediated suppression of THBS1 expression could promote cell proliferation and survival [107].

KHSV miR-K5 has also been shown to promote cell survival and growth by directly silencing Bcl-2-associated transcription factor 1 (BCLAF1) expression which acts as a transcriptional repressor of the anti-apoptotic protein Bcl-2 [108]. The over-expression of miR-K5 in B cells and endothelial cells resulted in decreased BCLAF1 levels. The siRNA-mediated silencing of BCLAF1 promoted lytic replication of KHSV suggesting BCLAF1 plays a key role in mediating the latent-to-lytic switch [108]. The KHSV miR-K11 and -K6 have been shown to specifically target the basic region leucine zipper (bZIP) transcription factor MAF (musculoaponeurotic fibrosarcoma oncogene homolog) [77, 87, 109] Since MAF plays a critical role in the terminal differentiation of many cell types, the KHSV miRNA-mediated silencing of MAF may regulate the differentiation state of infected endothelial cells that may promote oncogenesis.

The EBV encoded miRNA, miR-BART5 has been shown to target the p53-regulated proapoptotic Bcl-2 family member PUMA (p53 up-regulated modulator of apoptosis) [110]. Consistent with this effect, the suppression of miR-BART5 function increased PUMA-mediated apoptosis in EBV infected nasopharyngeal carcinoma (NPC) cells. The EBV miR-BHRF1-3 has been shown to down-regulate the interferon (IFN)-inducible chemokine ligand CXCL11. Since T cell immunity plays a major role in the host control of EBV infection, the silencing of CXCL11 may help EBV infected cells avoid recognition by cytotoxic T lymphocytes [111].

Since many viruses target common biological pathway that are involved in cell cycle regulation, cellular proliferation and survival, and immune evasion, it is not surprising that multiple virally-encoded miRNAs can target the same cellular gene. As an example of this principle, it has been shown that the major histocompatibility complex class I-related chain B (MICB) has unique binding sites for three virally-encoded miRNAs, EBV miR-BART2, hCMV miR-UL112-1, and KHSV miR-K7 [112, 113]. MICB serves as a ligand for natural killer (NK) cells and CD8+T cytotoxic T cells. Therefore, the reduction of MICB on the cell surface helps to prevent the recognition and killing of infected cells allowing the virus to evade immune control.

5 Cancer and miRNAs

MiRNAs are important regulators of a variety of biological processes including cellular proliferation, apoptosis, differentiation and development. Due to the intimate link between miRNAs and these essential biological pathways, it has come as no surprise that alterations in miRNA expression levels can have pathogenic consequences. This is most apparent in the development of cancers and metastases where miRNAs have been shown to act as tumor suppressors, oncogenes, and promoters or inhibitors of metastasis [114–116]. Since miRNAs act through the sequence-specific silencing of target mRNAs, the role that the dysregulation of miRNAs plays in cancer will be determined by the specific miRNA target genes. That is, the over-expression of a specific miRNA due to amplification of the genomic region encoding the miRNA or loss of epigenetic silencing and the promotion of transcription of that miRNA which targets one or more tumor suppressor gene could promote tumorogenesis. On the other hand, the deletion or epigenetic silencing of miRNAs that suppress the expression of one or more oncogene would relieve repression of these genes resulting in cellular transformation. Other miRNA related mechanisms that could lead to oncogenic transformation are the incorporation of mutations in a miRNA sequence (in particular the 'seed' region) altering the specificity of the miRNA and as a consequence the potential target mRNAs. Alternatively, changes in the sequence of the 3′ UTR either through the incorporation of mutations, insertions, deletions or translocations could alter the miRNA-mediated regulation of specific genes. Rather than providing a compendium of the miRNAs that have been associated with cancer development, this article will focus on the mechanisms by which alteration in miRNA expression levels or target binding sites can result in oncogenic transformation.

The first examples that miRNAs may be playing a role in the pathogenesis of cancer came when Croce et al. (2002) showed that a chromosomal region that spanned the coding sites fior two miRNAs (miR-15a and -16-1) was frequently deleted in chronic lymphocytic leukemia (CLL) [117]. MiR-15a and -16-1 function as tumor suppressors by inhibiting the expression of the antiapoptotic factors, BCL-2 and MCL-1. In contrast, the miR-17-92 cluster encoding six miRNAs, miR-17, -18a, -19a, -19b-1, -20a, and -92-1, is located in a genomic region that is frequently amplified in lymphomas [118]. The miR-17-92 cluster were the first miRNAs to have oncogenic functions by targeting the expression of the tumor suppressors Bim, PTEN, and CDKNI.[ref] see croce review.

In addition, large scale alteration of the miRNA processing machinery, such as the loss of expression of the ribonucleases Dicer or Drosha, have been shown to promote cellular transformation and tumor formation *in vivo*. This has been demonstrated by the experimentally induced silencing of these factors, either shRNA-mediated inhibition of Dicer and Drosha or conditional knock out of Dicer1 expression, in mouse models of cancer development [119, 120]. The loss of Dicer and/or Drosha has been shown to inversely correlate with the severity of outcomes in cancers derived from the ovarian epithelium. Interestingly, Martello et al [121] found that miR-103/107 which are over-expressed in many metastatic human breast cancers targeted the silencing of Dicer. This silencing of Dicer led to a global decrease in mature miRNAs levels, while specific loss of miR-200 expression promoted the epithelial-to-mesenchymal transition (EMT) and the metastatic potential of cancer cells.

Most miRNAs are expressed from RNA polymerase II promoters and, as such, are susceptible to aberrant transcription factor activity that can either increase or decrease miRNA expression levels. For example, the miR-34 family of miRNAs, miR-34a, -34b, and -34c, are directly induced by the tumor suppressor p53. Examining different tumor samples, Chang et al [122] found a direct correlation between p53 and miR-34 expression levels. Chromatin immunoprecipitation (ChIP) experiments showed that p53 directly bound to the promoter region of miR-34 [123–125]. Both the miR-29 family of miRNAs and miR-15-16 miRNA cluster are positively regulated by p53 [122–125]. Conversely, the oncogene MYC has been shown to negatively regulate the expression of members of the let-7 (let-7a, -7c, -7d, -7f1 and -7g) and miR-29 (miR-29a, -29b, and -29c) family of miRNAs [126]. These miRNAs have been shown to function as tumor suppressors through the targeted silencing of Ras by the let-7 family and myeloid leukemia differentiation protein (MCL1), T cell leukemia/lymphoma protein 1 (TCL1), cyclin dependent kinase 6 (CDK6) and DNA methyltransferase 3a (DNMT3a) by the miR-29 family [127–131]. Therefore, MYC-mediated repression of expression of these miRNAs would relieve the repression on these genes and promote tumor formation. MYC has been shown by ChIP experiments to bind to conserved sequences in the promoters of these miRNAs [126].

Changes to the epigenetic signature of miRNA promoters has been shown to alter miRNA expression levels in cancer cells. "The silencing of 'structurally normal' miRNA genes by DNA promoter hypermethylation and/or histone hypoacetylation has been described in solid tumors and in haematological malignancies [132–135]. For example, miR-127 expression was downregulated in response to DNA hypermethylation in bladder cancers. The experimentally induced loss of methylation led to the derepression of miR-127 expression and the concomitant silencing of the miR-127 target oncogene B-cell lymphoma protein-6 (BCL-6) [132]. Conversely, promoter methylation was shown to contribute to the down-regulation of miRNA expression in chronic lymphocytic leukemia (CLL) [136]

Tumorogenesis arises from the accumulation of genetic and/or epigenetic variations that facilitate oncogenic transformation. These genetic variations can include insertion, deletions, translocations, amplifications and single nucleotide polymorphisms. Similar to protein coding genes, these sources of genetic variation can alter miRNA functionality. The disruption of miRNA target binding sites by single nucleotide polymorphisms could result in increased cancer susceptibility and initiation. To that end, Chin et al. [137] identified a SNP in the let-7 target site in the 3′UTR of K-RAS. This alteration was associated with increased risk for the development of non-small cell lung cancers among moderate smokers. Similarly, this mutation functions as a genetic marker for ovarian cancer risk [138, 139]. In a genome-wide study, Calin et al. found SNP in miRNA binding sites in key protein associated with breast cancer including BRCA1 (breast cancer protein 1), MDM2 (murine double minute 2), TGFBR1 (transforming growth factor, beta receptor 1), and XIAP (X-linked inhibitor of apoptosis protein) [140]. The role of the target SNP in altering miRNA-mediated protein expression was shown using luciferase 3′UTR reporter assays for BRCA1 and TGFBR1 [140]. Similarly, Liang et al. [141] identified SNP in miRNA binding sites that correlated with increased risk for ovarian cancer and poorer survival. In addition to these small changes in single bases that disrupt miRNA binding to its target sites, larger translocations can occur which alter all or part of the 3′UTR of a gene changing the miRNA-mediated regulatory networks that control the expression of specific genes. For example, Mayr et al. [142] found that translocation in the HMGA2 (High mobility group AT-hook 2) often replaced the C terminus of the protein and the 3′UTR. The endogenous 3′UTR of HMGA2 has seven predicted binding sites for let-7 and HMGA2 expression has been shown by several groups to inhibit HMGA2 expression levels. When the 3' UTR of HMGA2 is replaced by a translocation event, HMGA2 will no longer be under let-7 control. This loss of let-7 regulation of HMGA2 resulted in increased HMGA2 levels and enhanced anchorage independent growth as measured by soft agar colony assays.

Although a strong polyadenylation signal usually is located at the 3′ end of the 3′UTR, almost all genes have additional proximal polyadenylation signals in their 3′UTRs, with about half of human genes possessing alternate polyadenylation signals whose use is supported by transcriptome data [143]. The use of these APA signals can eliminate large segments of the 3′UTR of genes which will decrease the number of potential miRNA binding sites and, as such, alter the miRNA-mediated regulatory networks that control gene expression levels. The analysis of 3′UTR lengths in tumor cells shows a distinct preference for shorter 3′UTRs resulting in increased protein expression levels. Importantly, the over-expression of the shorter but not the full length isoform of the insulin-like growth factor 2 mRNA binding protein 1 (IGF2BP1/IMP-1) relieved repression induced by let-7 increasing the stability of the mRNA and protein expression and promoted oncogenic transformation of fibroblasts and human breast epithelial cell lines [143].

6 Conclusions

A variety of molecular mechanisms have been exploited by viruses and tumor cells to manipulate miRNA-mediated gene regulatory pathways to promote their pathogenesis. For viruses, this has allowed for them to control cellular pathways that are necessary for their replication including increasing cell survival, cellular proliferation, differentiation, and/or immune evasion. Many viruses, in particular members of the herpesvirus family, encode miRNAs that not only regulate their own genes but also cellular genes creating a cellular environment that is conducive for their replication. Cancer development requires the accumulation of genetic and epigenetic variation that promote oncogenic transformation. MiRNAs have been identified that act as tumor suppressors or oncogenes based upon their specific targets. Alteration of miRNA expression levels in tumor cells can be accomplished through a variety of mechanisms, including structural changes to specific genomic locations (amplifications, deletions and point mutations), changes in transcription levels (repression or activation), epigenetic alterations (promoter hypermethylation or hypomethylation, histone deacetylation), and perturbation of the miRNA biogenesis pathway. Understanding the role that miRNAs play in the pathogenesis of viral infections and tumor development could provide potential novel targets for therapeutic development.

References

1. Bartel DP (2004) MicroRNAs: genomics, biogenesis, mechanism, and function. Cell 116:281–297
2. Bartel DP (2009) MicroRNAs: target recognition and regulatory functions. Cell 136:215–233
3. Carthew RW, Sontheimer EJ (2009) Origins and Mechanisms of miRNAs and siRNAs. Cell 136:642–655
4. Lee RC, Feinbaum RL, Ambros V (1993) The C. elegans heterochronic gene lin-4 encodes small RNAs with antisense complementarity to lin-14. Cell 75:843–854
5. Wightman B, Ha I, Ruvkun G (1993) Posttranscriptional regulation of the heterochronic gene lin-14 by lin-4 mediates temporal pattern formation in C. elegans. Cell 75:855–862
6. Reinhart BJ et al (2000) The 21-nucleotide let-7 RNA regulates developmental timing in Caenorhabditis elegans. Nature 403:901–906
7. Pasquinelli AE et al (2000) Conservation of the sequence and temporal expression of let-7 heterochronic regulatory RNA. Nature 408:86–89
8. Molnar A, Schwach F, Studholme DJ, Thuenemann EC, Baulcombe DC (2007) miRNAs control gene expression in the single-cell alga Chlamydomonas reinhardtii. Nature 447: 1126–1129
9. Grimson A et al (2008) Early origins and evolution of microRNAs and Piwi-interacting RNAs in animals. Nature 455:1193–1197
10. Kim VN, Nam, JW (2006) Genomics of microRNA. Trends Genet 22:165–173
11. Schanen BC, Li X Transcriptional regulation of mammalian miRNA genes. Genomics 97: 1–6
12. Lee Y et al (2004) MicroRNA genes are transcribed by RNA polymerase II. EMBO J 23:4051–4060
13. Cai X, Hagedorn CH, Cullen BR (2004) Human microRNAs are processed from capped, polyadenylated transcripts that can also function as mRNAs. RNA 10:1957–1966
14. Chen JF et al (2006) The role of microRNA-1 and microRNA-133 in skeletal muscle proliferation and differentiation. Nat Genet 38:228–233
15. Kim VN (2005) MicroRNA biogenesis: coordinated cropping and dicing. Nat Rev Mol Cell Biol 6:376–385
16. Kim VN, Han J, Siomi MC (2009) Biogenesis of small RNAs in animals. Nat Rev Mol Cell Biol 10:126–139
17. Han J et al (2004) The Drosha-DGCR8 complex in primary microRNA processing. Genes Dev 18:3016–3027
18. Han J et al (2006) Molecular basis for the recognition of primary microRNAs by the Drosha-DGCR8 complex. Cell 125:887–901
19. Gregory RI, Chendrimada TP, Shiekhattar R (2006) MicroRNA biogenesis: isolation and characterization of the microprocessor complex. Methods Mol Biol 342:33–47
20. Seitz H, Zamore PD (2006) Rethinking the microprocessor. Cell 125:827–829
21. Yi R et al (2009) DGCR8-dependent microRNA biogenesis is essential for skin development. Proc Natl Acad Sci U S A 106:498–502
22. Yeom KH, Lee Y, Han J, Suh MR, Kim VN (2006) Characterization of DGCR8/Pasha, the essential cofactor for Drosha in primary miRNA processing. Nucleic Acids Res 34:4622–4629
23. Landthaler M, Yalcin A, Tuschl T (2004) The human DiGeorge syndrome critical region gene 8 and Its D. melanogaster homolog are required for miRNA biogenesis. Curr Biol 14:2162–2167
24. Zeng Y, Cullen BR (2005) Efficient processing of primary microRNA hairpins by Drosha requires flanking nonstructured RNA sequences. J Biol Chem 280:27595–27603
25. Bohnsack MT, Czaplinski K, Gorlich D (2004) Exportin 5 is a RanGTP-dependent dsRNA-binding protein that mediates nuclear export of pre-miRNAs. RNA 10:185–191
26. Lund E, Guttinger S, Calado A, Dahlberg JE, Kutay U (2004) Nuclear export of microRNA precursors. Science 303:95–98

27. Yi R, Qin Y, Macara IG, Cullen BR (2003) Exportin-5 mediates the nuclear export of pre-microRNAs and short hairpin RNAs. Genes Dev 17:3011–3016
28. Grishok A et al (2001) Genes and mechanisms related to RNA interference regulate expression of the small temporal RNAs that control C. elegans developmental timing. Cell 106: 23–34
29. Hutvagner G et al (2001) A cellular function for the RNA-interference enzyme Dicer in the maturation of the let-7 small temporal RNA. Science 293:834–838
30. Ketting RF et al (2001) Dicer functions in RNA interference and in synthesis of small RNA involved in developmental timing in C. elegans. Genes Dev 15:2654–2659
31. Khvorova A, Reynolds A, Jayasena SD (2003) Functional siRNAs and miRNAs exhibit strand bias. Cell 115:209–216
32. Schwarz DS et al (2003) Asymmetry in the assembly of the RNAi enzyme complex. Cell 115:199–208
33. Bartel DP, Chen CZ (2004) Micromanagers of gene expression: the potentially widespread influence of metazoan microRNAs. Nat Rev Genet 5:396–400
34. Farh KK et al (2005) The widespread impact of mammalian MicroRNAs on mRNA repression and evolution. Science 310:1817–1821
35. Baek D et al (2008) The impact of microRNAs on protein output. Nature 455, 64–71
36. Friedman RC, Farh KK, Burge CB, Bartel DP (2009) Most mammalian mRNAs are conserved targets of microRNAs. Genome Res 19:92–105
37. Guo H, Ingolia NT, Weissman JS, Bartel DP Mammalian microRNAs predominantly act to decrease target mRNA levels. Nature 466:835–840
38. Shin C et al Expanding the microRNA targeting code: functional sites with centered pairing. Mol Cell 38:789–802
39. Selbach M et al (2008) Widespread changes in protein synthesis induced by microRNAs. Nature 455:58–63
40. Lewis BP, Burge CB, Bartel DP (2005) Conserved seed pairing, often flanked by adenosines, indicates that thousands of human genes are microRNA targets. Cell 120:15–20
41. Lewis BP, Shih IH, Jones-Rhoades MW, Bartel DP, Burge CB (2003) Prediction of mammalian microRNA targets. Cell 115:787–798
42. Lim LP, Glasner ME, Yekta S, Burge CB, Bartel DP (2003) Vertebrate microRNA genes. Science 299:1540
43. Lim LP et al (2003) The microRNAs of Caenorhabditis elegans. Genes Dev 17:991–1008
44. Lal A et al (2009) miR-24 Inhibits cell proliferation by targeting E2F2, MYC, and other cell-cycle genes via binding to "seedless" 3′UTR microRNA recognition elements. Mol Cell 35:610–625
45. Lim LP et al (2005) Microarray analysis shows that some microRNAs downregulate large numbers of target mRNAs. Nature 433:769–773
46. Lu R, Martin-Hernandez AM, Peart JR, Malcuit I, Baulcombe DC (2003) Virus-induced gene silencing in plants. Methods 30:296–303
47. Mueller S et al RNAi-mediated immunity provides strong protection against the negative-strand RNA vesicular stomatitis virus in Drosophila. Proc Natl Acad Sci U S A 107: 19390–19395
48. van Rij RP, Berezikov E (2009) Small RNAs and the control of transposons and viruses in Drosophila. Trends Microbiol 17:163–171
49. Saleh, M.C. et al (2009) Antiviral immunity in Drosophila requires systemic RNA interference spread. Nature 458:346–350
50. Aliyari R et al (2008) Mechanism of induction and suppression of antiviral immunity directed by virus-derived small RNAs in Drosophila. Cell Host Microbe 4:387–397
51. Flynt A, Liu N, Martin R, Lai EC (2009) Dicing of viral replication intermediates during silencing of latent Drosophila viruses. Proc Natl Acad Sci U S A 106:5270–5275
52. Kim J et al (2004) Identification of many microRNAs that copurify with polyribosomes in mammalian neurons. Proc Natl Acad Sci U S A 101:360–365

53. Grad Y et al (2003) Computational and experimental identification of C. elegans microRNAs. Mol Cell 11:1253–1263
54. Adai A et al (2005) Computational prediction of miRNAs in Arabidopsis thaliana. Genome Res 15:78–91
55. Li SC et al Identification of homologous microRNAs in 56 animal genomes. Genomics 96: 1–9
56. Farazi TA, Juranek SA, Tuschl T (2008) The growing catalog of small RNAs and their association with distinct Argonaute/Piwi family members. Development 135:1201–1214
57. Hafner M et al (2008) Identification of microRNAs and other small regulatory RNAs using cDNA library sequencing. Methods 44:3–12
58. Landgraf P et al (2007) A mammalian microRNA expression atlas based on small RNA library sequencing. Cell 129:1401–1414
59. Pfeffer S et al (2005) Identification of microRNAs of the herpesvirus family. Nat Methods 2:269–276
60. Pfeffer S et al (2004) Identification of virus-encoded microRNAs. Science 304:734–736
61. Cai X, Cullen BR (2006) Transcriptional origin of Kaposi's sarcoma-associated herpesvirus microRNAs. J Virol 80:2234–2242
62. Cai X et al (2005) Kaposi's sarcoma-associated herpesvirus expresses an array of viral microRNAs in latently infected cells. Proc Natl Acad Sci U S A 102:5570–5575
63. Cai X et al (2006) Epstein-Barr virus microRNAs are evolutionarily conserved and differentially expressed. PLoS Pathog 2:e23
64. Xing L, Kieff E (2007) Epstein-Barr virus BHRF1 micro- and stable RNAs during latency III and after induction of replication. J Virol 81:9967–9975
65. Samols MA, Hu J, Skalsky RL, Renne R (2005) Cloning and identification of a microRNA cluster within the latency-associated region of Kaposi's sarcoma-associated herpesvirus. J Virol 79:9301–9305
66. Zhu JY et al (2009) Identification of novel Epstein-Barr virus microRNA genes from nasopharyngeal carcinomas. J Virol 83:3333–3341
67. Grundhoff A, Sullivan CS, Ganem D (2006) A combined computational and microarray-based approach identifies novel microRNAs encoded by human gamma-herpesviruses. RNA 12:733–750
68. Umbach JL et al (2008) MicroRNAs expressed by herpes simplex virus 1 during latent infection regulate viral mRNAs. Nature 454:780–783
69. Umbach JL, Nagel MA, Cohrs RJ, Gilden DH, Cullen BR (2009) Analysis of human alphaherpesvirus microRNA expression in latently infected human trigeminal ganglia. J Virol 83:10677–10683
70. Cui C et al (2006) Prediction and identification of herpes simplex virus 1-encoded microRNAs. J Virol 80:5499–5508
71. Tang S, Patel A, Krause PR (2009) Novel less-abundant viral microRNAs encoded by herpes simplex virus 2 latency-associated transcript and their roles in regulating ICP34.5 and ICP0 mRNAs. J Virol 83:1433–1442
72. Tang S et al (2008) An acutely and latently expressed herpes simplex virus 2 viral microRNA inhibits expression of ICP34.5, a viral neurovirulence factor. Proc Natl Acad Sci U S A 105:10931–10936
73. Umbach JL et al Identification of viral microRNAs expressed in human sacral ganglia latently infected with herpes simplex virus 2. J Virol 84:1189–1192
74. Dunn W et al (2005) Human cytomegalovirus expresses novel microRNAs during productive viral infection. Cell Microbiol 7:1684–1695
75. Grey F et al (2005) Identification and characterization of human cytomegalovirus-encoded microRNAs. J Virol 79:12095–12099
76. Walz N, Christalla T, Tessmer U, Grundhoff A A global analysis of evolutionary conservation among known and predicted gammaherpesvirus microRNAs. J Virol 84:716–728
77. Gottwein E, Cai X, Cullen BR (2006) Expression and function of microRNAs encoded by Kaposi's sarcoma-associated herpesvirus. Cold Spring Harb Symp Quant Biol 71:357–364

78. Ahmed M, Lock M, Miller CG, Fraser NW (2002) Regions of the herpes simplex virus type 1 latency-associated transcript that protect cells from apoptosis *in vitro* and protect neuronal cells *in vivo*. J Virol 76:717–729
79. Gupta A, Gartner JJ, Sethupathy P, Hatzigeorgiou AG, Fraser NW (2006) Anti-apoptotic function of a microRNA encoded by the HSV-1 latency-associated transcript. Nature 442:82–85
80. Jones C (2003) Herpes simplex virus type 1 and bovine herpesvirus 1 latency. Clin Microbiol Rev 16:79–95
81. Inman M et al (2001) Region of herpes simplex virus type 1 latency-associated transcript sufficient for wild-type spontaneous reactivation promotes cell survival in tissue culture. J Virol 75:3636–3646
82. Bogerd HP et al A mammalian herpesvirus uses noncanonical expression and processing mechanisms to generate viral MicroRNAs. Mol Cell 37:135–142
83. Diebel KW, Smith AL, van Dyk LF Mature and functional viral miRNAs transcribed from novel RNA polymerase III promoters. RNA 16:170–185
84. Rodriguez A et al (2007) Requirement of bic/microRNA-155 for normal immune function. Science 316:608–611
85. Thai TH et al (2007) Regulation of the germinal center response by microRNA-155. Science 316:604–608
86. Burnside J et al (2006) Marek's disease virus encodes MicroRNAs that map to meq and the latency-associated transcript. J Virol 80:8778–8786
87. Skalsky RL et al (2007) Kaposi's sarcoma-associated herpesvirus encodes an ortholog of miR-155. J Virol 81:12836–12845
88. Bolisetty MT, Dy G, Tam W, Beemon KL (2009) Reticuloendotheliosis virus strain T induces miR-155, which targets JARID2 and promotes cell survival. J Virol 83:12009–12017
89. Yin Q et al (2008) MicroRNA-155 is an Epstein-Barr virus-induced gene that modulates Epstein-Barr virus-regulated gene expression pathways. J Virol 82:5295–5306
90. Xiao C, Rajewsky K (2009) MicroRNA control in the immune system: basic principles. Cell 136:26–36
91. Waidner LA et al (2009) MicroRNAs of Gallid and Meleagrid herpesviruses show generally conserved genomic locations and are virus-specific. Virology 388:128–136
92. Mathews MB, Shenk T (1991) Adenovirus virus-associated RNA and translation control. J Virol 65:5657–5662
93. Andersson MG et al (2005) Suppression of RNA interference by adenovirus virus-associated RNA. J Virol 79:9556–9565
94. Lu S, Cullen BR (2004) Adenovirus VA1 noncoding RNA can inhibit small interfering RNA and MicroRNA biogenesis. J Virol 78:12868–12876
95. Seo GJ, Chen CJ, Sullivan CS (2009) Merkel cell polyomavirus encodes a microRNA with the ability to autoregulate viral gene expression. Virology 383:183–187
96. Sullivan CS et al (2009) Murine Polyomavirus encodes a microRNA that cleaves early RNA transcripts but is not essential for experimental infection. Virology 387:157–167
97. Sullivan CS, Grundhoff AT, Tevethia S, Pipas JM, Ganem D (2005) SV40-encoded microRNAs regulate viral gene expression and reduce susceptibility to cytotoxic T cells. Nature 435:682–686
98. Cantalupo P et al (2005) Complete nucleotide sequence of polyomavirus SA12. J Virol 79:13094–13104
99. Grimson A et al (2007) MicroRNA targeting specificity in mammals: determinants beyond seed pairing. Mol Cell 27:91–105
100. Seo GJ, Fink LH, O'Hara B, Atwood WJ, Sullivan CS (2008) Evolutionarily conserved function of a viral microRNA. J Virol 82:9823–9828
101. Barth S et al (2008) Epstein-Barr virus-encoded microRNA miR-BART2 down-regulates the viral DNA polymerase BALF5. Nucleic Acids Res 36:666–675
102. Cullen BR (2009) Viral and cellular messenger RNA targets of viral microRNAs. Nature 457:421–425

103. Murphy E, Vanicek J, Robins H, Shenk T, Levine AJ (2008) Suppression of immediate-early viral gene expression by herpesvirus-coded microRNAs: implications for latency. Proc Natl Acad Sci U S A 105:5453–5458
104. Grey F, Nelson J (2008) Identification and function of human cytomegalovirus microRNAs. J Clin Virol 41:186–191
105. Grey F, Meyers H, White EA, Spector DH, Nelson J (2007) A human cytomegalovirus-encoded microRNA regulates expression of multiple viral genes involved in replication. PLoS Pathog 3:e163
106. Bellare P, Ganem D (2009) Regulation of KSHV lytic switch protein expression by a virus-encoded microRNA: an evolutionary adaptation that fine-tunes lytic reactivation. Cell Host Microbe 6:570–575
107. Samols MA et al (2007) Identification of cellular genes targeted by KSHV-encoded microRNAs. PLoS Pathog 3:e65
108. Ziegelbauer JM, Sullivan CS, Ganem D (2009) Tandem array-based expression screens identify host mRNA targets of virus-encoded microRNAs. Nat Genet 41:130–134
109. Gottwein E et al (2007) A viral microRNA functions as an orthologue of cellular miR-155. Nature 450:1096–1099
110. Choy EY et al (2008) An Epstein-Barr virus-encoded microRNA targets PUMA to promote host cell survival. J Exp Med 205:2551–2560
111. Xia T et al (2008) EBV microRNAs in primary lymphomas and targeting of CXCL-11 by ebv-mir-BHRF1-3. Cancer Res 68:1436–1442
112. Nachmani D, Stern-Ginossar N, Sarid R, Mandelboim O (2009) Diverse herpesvirus microRNAs target the stress-induced immune ligand MICB to escape recognition by natural killer cells. Cell Host Microbe 5:376–385
113. Stern-Ginossar N et al (2007) Host immune system gene targeting by a viral miRNA. Science 317:376–381
114. Dykxhoorn DM (2009) RNA interference as an anticancer therapy: a patent perspective. Expert Opin Ther Pat 19:475–491
115. Dykxhoorn DM, Chowdhury D, Lieberman J (2008) RNA interference and cancer: endogenous pathways and therapeutic approaches. Adv Exp Med Biol 615:299–329
116. Dykxhoorn DM MicroRNAs and metastasis: little RNAs go a long way. Cancer Res 70:6401–6406
117. Calin GA et al (2002) Frequent deletions and down-regulation of micro- RNA genes miR15 and miR16 at 13q14 in chronic lymphocytic leukemia. Proc Natl Acad Sci U S A 99:15524–15529
118. Diosdado B et al (2009) MiR-17–92 cluster is associated with 13q gain and c-myc expression during colorectal adenoma to adenocarcinoma progression. Br J Cancer 101:707–714
119. Kumar MS et al (2009) Dicer1 functions as a haploinsufficient tumor suppressor. Genes Dev 23:2700–2704
120. Merritt WM et al (2008) Dicer, Drosha, and outcomes in patients with ovarian cancer. N Engl J Med 359:2641–2650
121. Martello G et al (2010) A MicroRNA targeting dicer for metastasis control. Cell 141:1195–1207
122. Chang TC et al (2007) Transactivation of miR-34a by p53 broadly influences gene expression and promotes apoptosis. Mol Cell 26:745–752
123. He X, He L, Hannon GJ (2007) The guardian's little helper: microRNAs in the p53 tumor suppressor network. Cancer Res 67:11099–11101
124. He L, He X, Lowe SW, Hannon GJ (2007) microRNAs join the p53 network–another piece in the tumour-suppression puzzle. Nat Rev Cancer 7:819–822
125. He L et al (2007) A microRNA component of the p53 tumour suppressor network. Nature 447:1130–1134
126. Chang TC et al (2008) Widespread microRNA repression by Myc contributes to tumorigenesis. Nat Genet 40:43–50
127. Johnson SM et al (2005) RAS is regulated by the let-7 microRNA family. Cell 120:635–647

128. Garzon R et al (2009) MicroRNA-29b induces global DNA hypomethylation and tumor suppressor gene reexpression in acute myeloid leukemia by targeting directly DNMT3A and 3B and indirectly DNMT1. Blood 113:6411–6418
129. Fabbri M et al (2007) MicroRNA-29 family reverts aberrant methylation in lung cancer by targeting DNA methyltransferases 3A and 3B. Proc Natl Acad Sci U S A 104:15805–15810
130. Mott JL, Kobayashi S, Bronk SF, Gores GJ (2007) mir-29 regulates Mcl-1 protein expression and apoptosis. Oncogene 26:6133–6140
131. Garzon R et al (2009) MicroRNA 29b functions in acute myeloid leukemia. Blood 114:5331–5341
132. Saito Y et al (2006) Specific activation of microRNA-127 with downregulation of the proto-oncogene BCL6 by chromatin-modifying drugs in human cancer cells. Cancer Cell 9:435–443
133. Saito Y, Jones PA (2006) Epigenetic activation of tumor suppressor microRNAs in human cancer cells. Cell Cycle 5:2220–2222
134. Lujambio A et al (2007) Genetic unmasking of an epigenetically silenced microRNA in human cancer cells. Cancer Res 67:1424–1429
135. Hackanson B et al (2008) Epigenetic modification of CCAAT/enhancer binding protein alpha expression in acute myeloid leukemia. Cancer Res 68:3142–3151
136. Pallasch CP et al (2010) Blood. 114:3255
137. Chin LJ et al (2008) A SNP in a let-7 microRNA complementary site in the KRAS 3' untranslated region increases non-small cell lung cancer risk. Cancer Res 68:8535–8540
138. Keane FK, Ratner ES The KRAS-Variant Genetic Test As a Marker of Increased Risk of Ovarian Cancer. Rev Obstet Gynecol 3:118–121
139. Ratner E et al A KRAS-variant in ovarian cancer acts as a genetic marker of cancer risk. Cancer Res 70:6509–6515
140. Nicoloso MS et al Single-nucleotide polymorphisms inside microRNA target sites influence tumor susceptibility. Cancer Res 70:2789–2798
141. Liang D et al (2010) Genetic variants in MicroRNA biosynthesis pathways and binding sites modify ovarian cancer risk, survival, and treatment response. Cancer Res 70:9765–9776
142. Mayr C, Hemann MT, Bartel DP (2007) Disrupting the pairing between let-7 and Hmga2 enhances oncogenic transformation. Science 315:1576–1579
143. Mayr C, Bartel DP (2009) Widespread shortening of 3'UTRs by alternative cleavage and polyadenylation activates oncogenes in cancer cells. Cell 138:673–684
144. Du T, Zamore PD (2005) microPrimer: the biogenesis and function of microRNA. Development 132:4645–4652
145. Lee Y, Jeon K, Lee JT, Kim S, Kim VN (2002) MicroRNA maturation: stepwise processing and subcellular localization. EMBO J 21:4663–4670
146. Denli AM, Tops BB, Plasterk RH, Ketting RF, Hannon GJ (2004) Processing of primary microRNAs by the Microprocessor complex. Nature 432:231–235
147. Lee Y et al (2003) The nuclear RNase III Drosha initiates microRNA processing. Nature 425:415–419
148. Lee Y, Han J, Yeom KH, Jin H, Kim VN (2006) Drosha in primary microRNA processing. Cold Spring Harb Symp Quant Biol 71:51–57
149. Bernstein E, Caudy AA, Hammond SM, Hannon GJ (2001) Role for a bidentate ribonuclease in the initiation step of RNA interference. Nature 409:363–366
150. Chendrimada TP et al (2005) TRBP recruits the Dicer complex to Ago2 for microRNA processing and gene silencing. Nature 436:740–744
151. Matranga C, Tomari Y, Shin C, Bartel DP, Zamore PD (2005) Passenger-strand cleavage facilitates assembly of siRNA into Ago2-containing RNAi enzyme complexes. Cell 123:607–620
152. Hutvagner G, Simard MJ, Mello CC, Zamore PD (2004) Sequence-specific inhibition of small RNA function. PLoS Biol 2:E98
153. Hutvagner G, Zamore PD (2002) A microRNA in a multiple-turnover RNAi enzyme complex. Science 297:2056–2060

Oncogenic microRNAs in Cancer

Qian Liu, Nanjiang Zhou and Yin-Yuan Mo

Abstract MicroRNAs are a class of naturally occurring small non-coding RNAs that control gene expression as negative regulators at the post-transcriptional level. Since the discovery of microRNAs, the number of microRNAs has kept growing over the past years. To date over 1,000 human microRNA precursors have been identified and registered (www.miRBase.org). MicroRNAs exert their gene silencing function, usually by binding to the 3′-untranslated region (3′-UTR) of target genes through partial sequence homology, and thus, multiple protein-coding genes can be targeted by a given microRNA. Accordingly, microRNAs play a fundamental role in normal cell growth and disease processes. Particularly in cancer, microRNAs can function as either oncogenes or tumor suppressors. In this chapter, we will discuss our current understanding of a group of oncogenic microRNAs, focusing on *miR-21* and the *miR-17~92* cluster and their role in gene silencing, tumor growth and metastasis.

1 Introduction

It is well known that human genome transcribes a large amount of non-coding RNAs, however, it is not until recently that the functions of these non-coding RNAs have been investigated and appreciated [1]. MicroRNAs are among those non-coding RNAs that have been extensively characterized in recent years. Accumulating evidence indicates that microRNAs control gene expression through RNA interference (RNAi) pathway, leading to translational repression or degradation of mRNA in general [2–4]. Since their discovery in *C. elegans* over a decade ago [5–7], thousands

Q. Liu (✉) · N. Zhou
Department of Medical Microbiology, Immunology and Cell Biology,
Southern Illinois University School of Medicine, Springfield, IL 62794, USA

Y.-Y. Mo
Department of Medical Microbiology, Immunology and Cell Biology,
Southern Illinois University School of Medicine, 825 N. Rutledge,
PO Box 19626, Springfield, IL 62794, USA
e-mail: ymo@siumed.edu

S. Alahari (ed.), *MicroRNA in Cancer,*
DOI 10.1007/978-94-007-4655-8_5,

of microRNAs have been identified in a variety of organisms, including plants and animals through the genomics and bioinformatics effort (www.mirbase.org). Like protein-coding genes, microRNAs are transcribed as long primary transcripts (pri-microRNAs) in the nucleus, but they are produced after a serial of processes, including cutting, exporting and cutting by a number of enzymes including Drosha [8], exportin-5 [9], and the RNase III enzyme Dicer [10]. Finally a single-stranded mature microRNA is incorporated into the RNA-induced silencing complex (RISC) [11] to exert its silencing functions at the post-transcriptional level [12]. A unique feature of microRNAs is that it usually binds to the 3′-UTR of target genes through partial sequence homology, and thus, a single microRNA can have multiple targets [13, 14]. Together, microRNAs could regulate a large number of protein-coding genes. For example, as high as over 1/3 of the human genes could be under regulation of microRNAs, making them most abundant class of regulatory molecules [15].

In light of these functions, microRNAs are known to serve as master gene regulators. By silencing multiple targets involved in a particular pathway [16, 17], microRNAs can regulate various cellular processes, such as cell cycle [18, 19], proliferation [20], apoptosis [21, 22], differentiation [23, 24] and development [25]. Deregulation of microRNA expression is often associated with a variety of human diseases including cancer. For example, both *miR-15* and *miR-16* were downregulated in patients with B-cell chronic lymphocytic leukemia due to specific deletions on chromosome 13q14 [26]. Further studies indicate that those microRNAs are frequently located at cancer-associated chromosomal fragile sites. It has also been shown that the *let-7* family negatively regulates the Ras oncogene [27], and its therapeutic potential was demonstrated recently by reduction of lung tumors in mice after injecting *let-7* through intranasal route [28]. Although there are a large number of microRNAs involved in tumorigenesis, we will discuss a group of oncogenic microRNAs, focusing on *miR-21* and the *miR-17~92* cluster.

2 Role of Oncogenic microRNAs in Normal Cellular Processes

Overwhelming evidence suggest that microRNAs play a very broad role in a variety of cellular pathways, including development, differentiation, cell proliferation, apoptosis and even stem cell division and maintenance under physiological conditions. The role of *miR-21* and *miR-17~92* in these aspects has been extensively studied.

The finding that knockout of the microRNA processing enzyme Dicer leads to early arrest in development, accompanied by defects in proliferation of murine stem cells [29], provides a good indication of microRNAs as regulators for development. As a key factor for microRNA biogenesis, Dicer has also been implicated in controlling B lymphocyte development. Knockout of Dicer causes a developmental block at the pro- to pre-B cell transition. Specifically, *miR-17~92* is upregulated in

Dicer-deficient pro-B cells, whereas Bim, a target for *miR-17*, is upregulated [30]. This notion of microRNA-mediated regulation of development is further supported by the report that mice deficient in *miR-17~92* cluster have smaller embryos and immediate postnatal death of all animals and die shortly after birth with lung hypoplasia [31]. In contrast, transgenic mice expressing a high level of *miR-17~92* in lymphocytes develop lymphoproliferative disease and autoimmunity and die prematurely [32]. This is likely due to suppression of tumor suppressor PTEN and the proapoptotic protein Bim. These results highlight the importance of the *miR-17*-~92 cluster in development.

Similarly, *miR-21* has been shown to play a role in differentiation. For instance, *miR-21* is overexpressed in vascular walls and suppression of *miR-21* in vascular smooth muscle cells (VSMCs) and this results in decreased cell proliferation and increased apoptosis [33]. The *in vivo* studies further suggest a possible role of *miR-21* in neointimal lesion formation [33]. Of interest, ligand-specific SMAD proteins have been implicated in regulation of microRNA biogenesis involving *miR-21*, which is critical for control of the vascular smooth muscle cell phenotype [34]. Induction of a contractile phenotype in human vascular smooth muscle cells by TGF-beta and BMPs is mediated by *miR-21* [34]. The role of *miR-21* in mediating cardiomyocyte hypertrophy was demonstrated by Thum et al. [35]. Apparently, *miR-21* is able to regulate the ERK-MAP kinase signaling pathway in cardiac fibroblasts. In vivo silencing of *miR-21* experiments further suggest that *miR-21* can contribute to myocardial disease. However, a word of caution came from a recent report by Patrick et al regarding the role of *miR-21* in cardiac hypertrophy [36] because *miR-21*-null mice are normal. In addition, *miR-21* may function as a neural cell differentiation factor [37] and plays a role in regulation of myofibroblast differentiation. For example, the level of *miR-21* is often elevated in activated fibroblasts after treatment with TGF-β1 or conditioned medium from cancer cells. Interestingly, downregulation of *miR-21* with the *miR-21* inhibitor effectively inhibits TGF-β1-induced myofibroblast differentiation while upregulation of *miR-21* with a mimic significantly promotes myofibroblast differentiation [38].

Cell proliferation and apoptosis are normal cellular processes that can also be regulated by microRNAs. Early studies indicated that *miR-16* is able to specifically target Bcl-2, an anti-apoptotic protein [39], which may explain why miR-16 serves as a tumor suppressor. Overexpression of *miR-34a* promotes p53-mediated apoptosis [40, 41]. On the other hand, *miR-21* exerts an anti-apoptotic function because suppression of *miR-21* leads to increased apoptosis. *miR-21* can play a key role in suppressing IFN-induced apoptosis [42]. While IFN-induced apoptosis in PC3 cells is inhibited by *miR-21* overexpression, *miR-21* knockdown in DU145 cells enhances IFN-induced apoptosis [42].

Knockout of the *miR-17~92* cluster causes increased levels of the proapoptotic protein Bim as mentioned above; treatment with antagomir against *miR-17* abolishes the growth of MYCN-amplified and therapy-resistant neuroblastoma through upregulation of p21 and Bim, leading to activation of apoptosis [43]. Furthermore, a recent report suggests that the *miR-17~92* cluster is a novel target for p53-mediated transcriptional repression under hypoxia. In particular, while overexpression of *miR-17~92* cluster inhibits hypoxia-induced apoptosis, suppression of *miR-17*-5p and *miR-20a* sensitizes the cells to hypoxia-induced apoptosis [44]. In contrast, transgenic mice with overexpression of *miR-17* indicate that *miR-17* can reduce cell adhesion, migration and proliferation and thus these mice display overall growth retardation, smaller organs and greatly reduces hematopoietic cell lineages [45]. This is likely due to silencing of fibronectin and the fibronectin type-III domain containing 3A. It remains to be determined whether p53-mediated apoptosis is involved. Finally, *miR-17~*92 along with *miR-106b~25* has been implicated in modulating the transforming growth factor beta (TGF-β) [46], which serves as tumor suppressor under normal conditions. MicroRNAs can also have other cellular functions. For example, upregulation of *miR-21* promotes, while downregulation of *miR-21* attenuates the pro-fibrogenic activity of TGF-beta1 in fibroblasts [47], suggesting a role of *miR-21* in fibrotic lung diseases.

3 Oncogenic microRNAs vs Tumor Suppressive microRNAs

Up to date, numerous microRNAs have been shown to be associated with tumorigenesis, and they can be generally categorized into two groups, oncogenic microRNAs and tumor suppressive microRNAs. Oncogenic microRNAs include *miR-21*, and the *miR-17~92* cluster. On the other hand, tumor suppressor microRNAs are often downregulated in tumors and their expression may lead to suppression of tumor cell growth and invasion. They include *miR-15*, *miR-16*, *miR-145*, the *miR-34* cluster, and *miR-200* family. Based on available evidence, the number of the tumor suppressive microRNAs appears to be much larger than that of oncogenic microRNAs, which also is supported by reports that there is a general trend of downregulation of microRNAs in tumor specimens [48]. In general, these two groups of microRNAs either promote or suppress tumor cell growth. It is worth to mention a subgroup of microRNAs that may specifically affect invasion and metastasis, but have no effect on cell growth. For example, *miR-10b* is able to initiates robust invasion and metastasis in non-metastatic tumors by inhibiting translation of homeobox D10 mRNA, resulting in increased expression of a well-characterized pro-metastatic gene, RHOC [49]. However, this could be cell type specific. For example, we found that *miR-145* suppresses cell growth in MCF-7 and HCT-116 cells, but it has little effect on growth of the metastatic breast cancer cell line MDA-MB-231, and instead significantly suppresses its invasion and metastasis [50].

4 Oncogenic microRNA-Mediated Tumor Growth

Since microRNAs can serve as master gene regulators by silencing multiple protein-coding genes, it is not surprising that deregulation of microRNAs could disrupt normal cell growth and development, leading to a variety of disorders including

Table 1 Selective oncogenic microRNAs ant their targets

MicroRNA	Representative target	Reference
hsa-miR-21	PTEN	Meng et al. [63]
	TPM1	Zhu et al. [64]
	PDCD4	Asangani et al. [109]
hsa-miR-17~92	Bim	Ventura et al. [31]
hsa-miR-10b	HOXD4	Ma et al. [49]
hsa-miR-125b	p53	Shi et al. [98]
	DEICER	Klusmann [110]
hsa-miR-155	FOXO3a	Kong et al. [111]
hsa-miR-221/222	p27	Galardi et al. [112]
hsa-miR-373	CD44	Huang et al. [113]
hsa-miR-375	RASD1	de Souza Rocha Simonini et al. [114]
hsa-miR-504	p53	Hu et al. [115]

human cancer [12, 22, 51–53]. Oncogenic microRNAs are often upregulated in tumor specimens, and this upregulation is associated with downregulation of their corresponding tumor suppressors. Table 1 lists oncogenic microRNAs which have been relatively well characterized. Among them, *miR-21* and *miR-17~92* are probably the most studied oncogenic microRNAs.

There is overwhelming evidence that *miR-21* is upregulated in a variety of tumors including breast, prostate, colon, liver and lung cancer [54, 55]. More recently, upregulation of *miR-21* has also been reported in squamous cell carcinoma [56, 57] and cholangiocarcinoma [58]. A more comprehensive list of *miR-21* dysregulation can be found in the Human MiRNA & Disease Database (HMDD) site (http://202.38.126.151/hmdd/mirna/md/). As an oncogene, *miR-21* targets a number of genes involved in cell growth, proliferation and apoptosis. Ectopic expression of *miR-21* enhances, while suppression of *miR-21* inhibits cell growth, as demonstrated in various cell culture or animal models. For example, knockdown of *miR-21* in cultured glioblastoma cells triggers activation of caspases and leads to increased apoptotic cell death [59], which may explain why *miR-21* is often upregulated in the highly malignant glioblastoma. Similarly, experiments with an anti-*miR-21* approach showed that breast cancer MCF-7 cells transfected with anti-*miR-21* formed smaller tumors than control in female nude mice [60], suggesting that transient suppression of *miR-21* is sufficient to cause tumor growth inhibition. We have further identified Bcl-2 as a potential indirect target of *miR-21* because anti-*miR-21* causes downregulation of Bcl-2 [60].

Although over a thousand of putative targets are predicted by computer algorithms, based on Mirecords (http://mirecords.biolead.org/), there are 27 *miR-21* direct targets that are experimentally validated. DIANA Lab lists a total of nine entries (http://diana.cslab.ece.ntua.gr/tarbase/). The early identified targets include PTEN, TPM1, PDCD4, maspin, RECK and TIMP3, CDC25A. The number of experimentally validated *miR-21* targets is still growing. More recent studies have identified additional targets such as TIAM1 and PPARα [61, 62]. The first validated *miR-21* target is PTEN [63] in human hepatocellular cancer cell lines. Given the role of PTEN in suppression of tumor growth, identification of PTEN as a *miR-21* target may explain the significance of *miR-21* as an oncogene because as a repressor for PI3 K, suppression of pTEN would promote Akt-mediated cell growth and proliferation. We identified tropomyocin 1 (TPM1) as a direct target for *miR-21* by proteomic approach combined with luciferase reporter and western blot [64]. Both PTEN and TPM1 are known tumor suppressors. There are also other tumor suppressors such as PDCD4, maspin RECK and TIMP3 that were shown to be targets for *miR-21* [65–68]. Of considerable interest, *miR-21*, like many other microRNAs, is able to simultaneously silence multiple genes involved in cell growth and apoptosis.

Experiments with transgenic mice through the loss-of-function or gain-of-function approach revealed that overexpression of *miR-21* enhances tumorigenesis whereas deletion of *miR-21* partially protects against tumor formation in a lung cancer model [69]. The underlying mechanism may involve suppression of negative regulators of the Ras/MEK/ERK pathway and inhibition of apoptosis [69]. Interestingly, cell specific modulations of *miR-21* provide evidence that *miR-21* functions as an oncogene in development of lymphoma [70].

The *miR-17~92* cluster carries 6 microRNAs; several of them have been shown to silence genes involved in cell growth and apoptosis. For example, *miR-17* and *miR-20a* control the balance of cell death and proliferation driven by the proto-oncogene c-Myc [19]; *miR-17* may serve as an oncogene in lymphoma and lung cancer [71]. More recently, *miR-19* alone was shown to be sufficient to promote c-Myc-induced lymphomagenesis by repressing apoptosis through silencing PTEN [72]. By activation of the Akt-mTOR pathway, *miR-19* functionally antagonizes PTEN to promote cell survival. In another report, *miR-19a* and *miR-19b* were shown to be required and largely sufficient to recapitulate the oncogenic properties of the entire cluster [73]. Furthermore, *miR-19* is sufficient to promote leukemogenesis in Notch1-induced T-cell acute lymphoblastic leukemia *in vivo* [74]. These findings suggest that a single microRNA from the cluster can account for the observed oncogenic phenotype and highlight the importance of individual microRNAs in tumorigenesis. Finally, *miR-17~92* is frequently amplified in mixed lineage leukemia (MLL)-rearranged acute leukemias and it can significantly increase colony-forming capacity of normal mouse bone marrow progenitor cells, suggesting a key role of this cluster in the development of MLL-associated leukemias [75].

In addition to hematopoietic cancer, the *miR-17~92* cluster may also function as an oncogene in solid tumors. For example, *miR-17* may also serve as an oncogene in lung cancer [76]; *miR-17*-3p (equivalent to *miR-17**) is able to substantially suppress vimentin levels and restore the formation of polarized acini in laminin-rich extracellular matrix gel in prostate cancer cells [77]. *miR-17*-5p (equivalent to *miR-17*) up-regulates proliferation of hepatocellular carcinoma (HCC) cells by activating the p38 mitogen-activated protein kinase MAPK pathway and increases the phosphorylation of heat shock protein 27 (HSP27) [78]. Finally, both *miR-17*-5p and *miR-20a* of this cluster have been shown to play an oncogenic role in renal cell carcinoma [79]

5 Oncogenic microRNA-Mediated Invasion and Metastasis

Metastasis is a major cause of cancer related death and thus, how microRNAs affect tumor invasion and metastasis has been a hot topic in the microRNA research field. As discussed above, microRNAs such as *miR-10b* appears to affect metastasis without affecting tumor cell growth [49], while other microRNAs may regulate invasion and metastasis in addition to cell growth and proliferation. However, it remains to be determined as to whether this is a cell type specific phenomenon. For example, we have found that *miR-21* is able to promote tumor growth in the MCF-7 model [60]. However, in the metastatic breast cancer MDA-MB-231 model, *miR-21* has little effect on cell growth in cell culture, or primary tumor growth in xenograft animal model [66].

Since there is a positive correlation between *miR-21* expression and metastasis status [80–83], *miR-21* could be responsible for clinical tumor invasion and metastasis. This notion first came from the observation that *miR-21* expression correlates with metastasis status [80]. Subsequently, we and others [66, 67] demonstrated that *miR-21* has a potential role in mediating invasion and metastasis in part by targeting tumor suppressor genes in invasion. For example, we showed that in our xenograft breast cancer metastatic model, suppression of *miR-21* in MDA-MB-231 cells does not seem to affect cell growth *in vitro* or primary tumor growth, but it causes a significant reduction both *in vitro* invasion and *in vivo* lung metastasis [66]. This is likely due to silencing of multiple genes such as PDCD4, TPM1 and maspin, all of which have been well implicated in invasion and metastasis. Importantly, there is a significant negative correlation between *miR-21* and PDCD4 in matched breast tumor specimens, suggesting its clinical significance. Furthermore, there is an inverse relationship between *miR-21* and PDCD4 in ten colorectal cancer cell lines and 22 resected normal/tumor tissues [67]. Anti-*miR-21* can reduce intravasation and lung metastasis of RKO cells in a chicken-embryo-metastasis assay. These studies are consistent with the report that a poor prognosis of metastatic cancer patients is associated with high *miR-21* expression [81]. In glioblastoma, *miR-21* regulates multiple genes associated with apoptosis, migration, and invasiveness, including the RECK and TIMP3 genes, which are suppressors of malignancy and inhibitors of matrix metalloproteinases (MMPs) [84]. Thus, *miR-21* appears to contribute to glioma malignancy by downregulation of MMP inhibitors, which in turn activates MMPs and promotes invasiveness. A matrigel invasion assay showed that PDCD4 expression suppressed invasion, and siRNA-mediated PDCD4 loss was associated with increased invasive potential of oral carcinoma cells. Furthermore, *miR-21* levels are increased in PDCD4-negative tumors. PDCD4 expression may be down-regulated in oral squamous cell carcinoma by direct binding of *miR-21* to the 3′-UTR of PDCD4 mRNA [85]. Elevated levels of *miR-21* and *miR-31* also enhance cell motility and invasiveness of colon carcinoma cell lines, which may work through the TGF-β pathway [61].

Although the *miR-17~92* cluster plays an important role in hematopoietic cancer, several reports support the notion that it may be also important in solid tumors. However, its role in cell invasion and metastasis is somewhat controversial or could also be cell type specific. While the *miR-17~92* cluster usually plays an oncogenic role, it can also function as tumor suppressor in invasion and metastasis. Of particular interest, this *miR-17~92* cluster was recently shown to regulate cellular migration and invasion of nearby cells via heterotypic secreted signals in a breast cancer model [86]. For example, cell-conditioned medium from non-invasive breast cancer MCF-7 cells with overexpression of *miR-17~92* is able to inhibit migration and invasion of metastatic MDA-MB-231 cells; this appears to be through inhibiting secretion of a subset of cytokines, and suppressing plasminogen activation via inhibition of the secreted plasminogen activators [86]. Although this is an artificial system, it may be the first step toward understanding of how microRNAs may play a key role in invasion and metastasis through such communications in tumor microenvironments. In prostate cancer cells, *miR-17*–3p has been shown to suppress vimentin levels and restore the formation of polarized acini in laminin-rich extracellular matrix gel; and reduce migration and invasion [77].

On the other hand, a recent study indicates that expression of *miR-17*–5p correlates with pancreatic cancer progression and invasion in clinical specimens [87]. Furthermore, ectopic expression of *miR-17* precursor increases the number of invasive cells in pancreatic cancer cell lines [87]. Similarly, *miR-17*-5p was shown to be overexpressed in high-invasive MDA-MB-231 breast cancer cells compared to non-invasive MCF-7 breast cancer cells. Ectopic expression of *miR-17* promotes invasiveness of MCF-7 cells while down-regulation of *miR-17* suppresses the migration and invasion of MDA-MB-231 cells [88]. The role of *miR-17* as an oncogene in promoting invasion and metastasis was further supported by the finding that blockade of *miR-17* is able to decrease breast cancer cell invasion/migration *in vitro* and metastasis *in vivo* [89]. Upregulation of *miR-17*-5p increases the migration of heptoma cells [78].

6 Cell Type Specific Effect of microRNAs

Just like transcription factors, microRNA-mediated gene silencing can also be cell type specific or can be influenced by environmental stimuli. For example, one microRNA may specifically silences a set of targets in breast cancer, but the same microRNA may not necessarily be able to silence the same set of targets in lung cancer, suggesting the complexity of microRNA targeting. In support of this notion, we have recently found that *miR-101* has inhibitory effect on cell growth in estrogen containing medium, however it promotes cell growth in estrogen free medium [90]. In this regard, although *miR-21* is well known as an oncogene in many types of cancers, *miR-21* was shown to suppress Cdc25A in colon cancer cells [91]. Cdc25A is a member of Cdc25 family proteins, and correlates with more aggressive disease and poor prognosis in some cancers and leads to genetic instability in mice.

In consistent with this finding, *miR-21* is underexpressed in a subset of Cdc25A-overexpressing colon cancers [91].

miR-17 is a good example of microRNAs that may serve as oncogenes or tumor suppressors depending the cellular content. As discussed early, it is evident that the *miR-17~92* cluster serves as an oncogene in hematopoietic cancer. However, in solid tumors its role is controversial. For example, there are several reports of *miR-17* as a tumor suppressor in breast cancer. On one hand, *miR-17*-5p regulates breast cancer cell proliferation by silencing AIB1 [92]; overexpression of *miR-17~92* is able to inhibit migration and invasion of metastatic breast cancer MDA-MB-231 cells through heterotypic secreted signals [86]. On the other hand, *miR-17*-5p seems to promote migration and invasiveness of non-invasive breast cancer cells by suppressing HBP1 and subsequent activation of Wnt/β-catenin [88]. Evidently, more work is needed to reconcile different phenotypes caused by *miR-17*. This cell type specific phenomenon may in part be explained by a recent finding that *miR-17~92* is a cell cycle regulated locus and *miR-17*-5p acts specifically at the G1/S-phase cell cycle boundary, by targeting multiple genes involved in the transition between these phases. While both pro- and anti-proliferative genes are targeted by *miR-17*–5p, pro-proliferative mRNAs are specifically up-regulated by secondary and/or tertiary effects [93].

7 Oncogenic microRNA-Mediated Resistance to Chemotherapy and Radiation Therapy

MicroRNAs can play a role not only in tumorigenesis, but also in other aspects, in particular chemo-resistance or radiation resistance. For example, we have previously shown that *miR-21* can sensitize MCF-7 cells to topotecan [60]. Similarly, *miR-21* inhibition increases the sensitivity of malignant cholangiocytes to gemcitabine [94]. A recent report suggested that *miR-21* expression status may serve as a predictive biomarker for treatment outcome. For example, low *miR-21* expression is correlated with longer overall as well as disease-free survival. Importantly, suppression of *miR-21* by anti-*miR-21* enhances the chemosensitivity of pancreatic ductal adenocarcinoma cells, raising a possibility for adjuvant therapy using anti-*miR-21* [95]. Anti-*miR-21* oligonucleotide has been shown to sensitize K562 and HL60 cells to chemotherapy agents [96, 97]. For instance, anti-*miR-21* along with Ara-C increases the sensitivity of HL60 cells to Ara-C and promotes Ara-C-induced apoptosis through upregulation of *miR-21* direct target PDCD4 [97]. Furthermore, overexpression of *miR-21* could significantly reduce Temozolomide-induced apoptosis in human glioblastoma U87MG cells, which may involve a shift in Bax/Bcl-2 ratio and change in caspase-3 activity [98]. Of interest, serum *miR-21* levels are elevated in hormone-refractory prostate cancer (HRPC) patients, especially in those resistant to docetaxel-based chemotherapy [99].

MicroRNA-mediated chemo-resistance is not limited to solid tumors. *miR-21* was shown to be one of the microRNAs associated with fludarabine resistance in chronic lymphocytic leukemia (CLL) patients. *In vitro* cell culture experiments revealed that inhibition of *miR-21* by antagamir induces a significant increase in caspase activity in fludarabine-treated p53-mutant MEG-01 cells, suggesting the involvement of *miR-21* in the establishment of fludarabine resistance [100].

Information about *miR-17* in this aspect is limited. However, in primary neuroblastoma, suppression of *miR-17* with antagomir treatment can significantly inhibits tumor growth *in vivo*, specifically in therapy-resistant neuroblastoma cells [101].

8 Oncogenic microRNAs as a Biomarker for Cancer Diagnosis and Prognosis, and as Therapeutic Targets

Since microRNAs play a fundamental role in regulation of diverse cellular pathways, deregulation of microRNA expression is often detected in tumor specimens and thus, a unique set of microRNAs (or microRNA signatures) due to their alteration would serve as markers for cancer diagnosis and prognosis. For example, early studies showed a general downregulation of a number of microRNAs in tumors compared with normal tissues in multiple human cancers [48]. Such microRNA signatures have also been reported in many other types of cancers. As an oncogenic microRNA, *miR-21* is upregulated in many types of solid tumors, and is often associated with clinical features. These include colorectal carcinomas [102] [54] [103]. In particular, higher levels of *miR-21* were detected in adenomas and in tumors with more advanced TNM staging and were associated with poor survival, underscoring the importance of *miR-21* as prognostic factor for colorectal carcinomas, including response to chemotherapy. For example, upregulation of *miR-21* is associated with clinical features in non-small cell lung cancer (NSCLC) [104].

Of considerable interest, recent evidence indicates that microRNAs can be stably present in the circulating system and some of these circulating microRNAs originate from tumors [105], suggesting that blood/serum is a potential source for detection of cancer-associated microRNAs. Since the circulating microRNAs appear to be quite resistant to RNase digestion and multiple cycles of freeze-thaw presumably this greatly simplify sample handling and thus, they may be suited for multiplexing assays. Apparently, this is a new area; more work is needed in terms of detection sensitivity, specificity and reliability.

Oncogenic microRNAs have also been shown to serve as good therapeutic targets and in particular targeting oncogenic microRNAs seem to therapeutically practical. A recent important discovery that the level of *miR-21* alone can determine the fate of cancer cells [70] provides a strong rationale of targeting microRNAs as a promising approach for cancer treatment. Since *miR-21* is often upregulated in tumors, this may provide an "opportunity window" for specifically targeting tumor cells. In this regard, there are many forms of anti-sense oligonucleotides (ASOs) available for this purpose. Basically, they are products with various modifications which could increase their stability [106] and potency. For example, 3′-cholesterol modified ASOs have enhanced potency and are stable for at least 7 days from a single transfection. Another important modification is locked nucleic acid (LNA) DNA oligos [107]. One obvious advantage of LNA oligos over other modified antisense oligos is DNA-based, which is presumably more stable. In terms of inhibition ability, both modified anti-*miR-21* RNA oligo and LNA anti-*miR-21* oligo seem to function well [59].

Two groups reported that suppression of *miR-21* by ASOs can significantly decrease the cell growth in cholangiocarcinomas and glioma cells, respectively [59, 94]. We demonstrated that anti-*miR-21* is effective in suppression of *miR-21* in breast cancer cell lines MCF-7 and MBA-MD-231. Of interest, this suppression can be measured by real time RT-PCR, which reveals a significant reduction of the endogenous mature *miR-21*. Importantly, this suppression of *miR-21* causes not only cell growth inhibition in cell culture model, but also suppression of tumor growth and metastasis in an animal model [60, 66].

Although *miR-21* is often downregulated in hematopoietic cancer [108], surprisingly a recent report indicates that *miR-21* serves an oncogene and suppression of *miR-21* can lead to cure of cancer [70], supporting the notion of oncomiR addiction. For example, overexpression of *miR-21* leads to a pre-B malignant lymphoid-like phenotype. In contrast, suppression of *miR-21* causes tumor regression, highlighting a possibility that pharmacological inactivation of oncogenic miRNAs alone may provide effective way to treat human cancers.

9 Summary

Oncogenic microRNAs are a group of microRNAs that are capable of promoting tumor cell growth, invasion and metastasis. We use *miR-21* and *miR-17~92* as examples to illustrate how oncogenic microRNAs impact these aspects. Evidently, the key to the oncogenic features of these microRNAs is their ability to specifically target a spectrum of protein-coding genes in a given cellular content. Because of this feature, oncogenic microRNAs is a relative term, as discussed about, and thus, oncogenic microRNAs may also reveal tumor suppressive features under different circumstances. Therefore, a better understanding of molecular mechanism underlying microRNA-mediated gene expression will help design better strategy for cancer treatment. Their immediate clinical implications will be very much dependent on how these microRNAs can be served as biomarkers or therapeutic targets.

References

1. Turner AM, Morris KV (2010) Controlling transcription with noncoding RNAs in mammalian cells. Biotechniques 48:ix–xvi
2. Pillai RS (2005) MicroRNA function: multiple mechanisms for a tiny RNA? Rna 11:1753–1761

3. Zamore PD, Haley B (2005) Ribo-gnome: the big world of small RNAs. Science 309:1519–1524
4. Bartel DP (2004) MicroRNAs: genomics, biogenesis, mechanism, and function. Cell 116:281–297
5. Lee RC, Feinbaum RL, Ambros V (1993) The C. elegans heterochronic gene lin-4 encodes small RNAs with antisense complementarity to lin-14. Cell 75:843–854
6. Wightman B, Ha I, Ruvkun G (1993) Posttranscriptional regulation of the heterochronic gene lin-14 by lin-4 mediates temporal pattern formation in C. elegans. Cell 75:855–862
7. Reinhart BJ, Slack FJ, Basson M, Pasquinelli AE, Bettinger JC, Rougvie AE, Horvitz HR, Ruvkun G (2000) The 21-nucleotide let-7 RNA regulates developmental timing in caenorhabditis elegans. Nature 403:901–906
8. Kim VN (2005) MicroRNA biogenesis: coordinated cropping and dicing. Nat Rev Mol Cell Biol 6:376–385
9. Yi R, Qin Y, Macara IG, Cullen BR (2003) Exportin-5 mediates the nuclear export of pre-microRNAs and short hairpin RNAs. Genes Dev 17:3011–3016
10. Carmell MA, Hannon GJ (2004) RNase III enzymes and the initiation of gene silencing. Nat Struct Mol Biol 11:214–218
11. Du T, Zamore PD (2005) microPrimer: the biogenesis and function of microRNA. Development 132:4645–4652
12. Esquela-Kerscher A, Slack FJ (2006) Oncomirs—microRNAs with a role in cancer. Nat Rev Cancer 6:259–269
13. Brennecke J, Stark A, Russell, RB, Cohen SM (2005) Principles of microRNA-target recognition. PLoS Biol 3:e85
14. Friedman RC, Farh KK, Burge CB, Bartel DP (2009) Most mammalian mRNAs are conserved targets of microRNAs. Genome Res 19:92–105
15. Lewis BP, Burge CB, Bartel DP (2005) Conserved seed pairing, often flanked by adenosines, indicates that thousands of human genes are microRNA targets. Cell 120:15–20
16. Obernosterer G, Tafer H, Martinez J (2008) Target site effects in the RNA interference and microRNA pathways. Biochem Soc Trans 36:1216–1219
17. Vella MC, Reinert K, Slack FJ (2004) Architecture of a validated microRNA: target interaction. Chem Biol 11:1619–1623
18. Hatfield SD, Shcherbata HR, Fischer KA, Nakahara K, Carthew RW, Ruohola-Baker H (2005) Stem cell division is regulated by the microRNA pathway. Nature 435:974–978
19. O'Donnell KA, Wentzel EA, Zeller KI, Dang CV, Mendell JT (2005) c-Myc-regulated microRNAs modulate E2F1 expression. Nature 435:839–843
20. Ambros V (2004) The functions of animal microRNAs. Nature 431:350–355
21. Tong AW, Nemunaitis J (2008) Modulation of miRNA activity in human cancer: a new paradigm for cancer gene therapy? Cancer Gene Ther 15:341–355
22. Croce CM, Calin GA (2005) miRNAs, cancer, and stem cell division. Cell 122:6–7
23. Moss EG (2007) Heterochronic genes and the nature of developmental time. Curr Biol 17:R425–434
24. Martinez-Climent JA, Andreu EJ, Prosper F (2006) Somatic stem cells and the origin of cancer. Clin Transl Oncol 8:647–663
25. Ambros V (2003) MicroRNA pathways in flies and worms: growth, death, fat, stress, and timing. Cell 113:673–676
26. Calin GA, Croce CM (2006) Genomics of chronic lymphocytic leukemia microRNAs as new players with clinical significance. Semin Oncol 33:167–173
27. Johnson SM, Grosshans H, Shingara J, Byrom M, Jarvis R, Cheng A, Labourier E, Reinert KL, Brown D, Slack FJ (2005) RAS is regulated by the let-7 microRNA family. Cell 120:635–647
28. Esquela-Kerscher A, Trang P, Wiggins JF, Patrawala L, Cheng A, Ford L, Weidhaas JB, Brown D, Bader AG, Slack FJ (2008) The let-7 microRNA reduces tumor growth in mouse models of lung cancer. Cell Cycle 7:759–764

29. Murchison EP, Partridge JF, Tam OH, Cheloufi S, Hannon GJ (2005) Characterization of dicer-deficient murine embryonic stem cells. Proc Natl Acad Sci U S A 102:12135–12140
30. Koralov SB, Muljo SA, Galler GR, Krek A, Chakraborty T, Kanellopoulou C, Jensen K, Cobb BS, Merkenschlager M, Rajewsky N, Rajewsky K (2008) Dicer ablation affects antibody diversity and cell survival in the B lymphocyte lineage. Cell 132:860–874
31. Ventura A, Young AG, Winslow MM, Lintault L, Meissner A, Erkeland SJ, Newman J, Bronson RT, Crowley D, Stone JR, Jaenisch R, Sharp PA, Jacks T (2008) Targeted deletion reveals essential and overlapping functions of the miR-17 through 92 family of miRNA clusters. Cell 132:875–886
32. Xiao C, Srinivasan L, Calado DP, Patterson HC, Zhang B, Wang J, Henderson JM, Kutok JL, Rajewsky K (2008) Lymphoproliferative disease and autoimmunity in mice with increased miR-17-92 expression in lymphocytes. Nat Immunol 9:405–414
33. Ji R, Cheng Y, Yue J, Yang J, Liu X, Chen H, Dean DB, Zhang C (2007) MicroRNA expression signature and antisense-mediated depletion reveal an essential role of MicroRNA in vascular neointimal lesion formation. Circ Res 100:1579–1588
34. Davis BN, Hilyard AC, Lagna G, Hata A (2008) SMAD proteins control DROSHA-mediated microRNA maturation. Nature 454:56–61
35. Thum T, Gross C, Fiedler J, Fischer T, Kissler S, Bussen M, Galuppo P, Just S, Rottbauer W, Frantz S, Castoldi M, Soutschek J, Koteliansky V, Rosenwald A, Basson MA, Licht JD, Pena JT, Rouhanifard SH, Muckenthaler MU, Tuschl T, Martin GR, Bauersachs J, Engelhardt S (2008) MicroRNA-21 contributes to myocardial disease by stimulating MAP kinase signalling in fibroblasts. Nature 456:980–984
36. Patrick DM, Montgomery RL, Qi X, Obad S, Kauppinen S, Hill JA, van Rooij E, Olson E N (2010) Stress-dependent cardiac remodeling occurs in the absence of microRNA-21 in mice. J Clin Invest 120(11):3912–3916. doi: 10.1172/JCI43604. Epub 2010 Oct 18
37. Singh SK, Kagalwala MN, Parker-Thornburg J, Adams H, Majumder S (2008) REST maintains self-renewal and pluripotency of embryonic stem cells. Nature 453(7192):223–227 Epub 2008 Mar 23
38. Yao Q, Cao S, Li C, Mengesha A, Kong B, Wei M (2011) Micro-RNA-21 regulates TGF-beta-induced myofibroblast differentiation by targeting PDCD4 in tumor-stroma interaction. Int J Cancer 128(8):1783–1792. doi: 10.1002/ijc.25506
39. Calin GA, Dumitru CD, Shimizu M, Bichi R, Zupo S, Noch E, Aldler H, Rattan S, Keating M, Rai K, Rassenti L, Kipps T, Negrini M, Bullrich F, Croce CM (2002) Frequent deletions and down-regulation of micro- RNA genes miR15 and miR16 at 13q14 in chronic lymphocytic leukemia. Proc Natl Acad Sci U S A 99:15524–15529
40. Chang TC, Wentzel EA, Kent OA, Ramachandran K, Mullendore M, Lee KH, Feldmann G, Yamakuchi M, Ferlito M, Lowenstein CJ, Arking DE, Beer MA, Maitra A, Mendell JT (2007) Transactivation of miR-34a by p53 broadly influences gene expression and promotes apoptosis. Mol Cell 26:745–752
41. Raver-Shapira N, Marciano E, Meiri E, Spector Y, Rosenfeld N, Moskovits N, Bentwich Z, Oren M (2007) Transcriptional activation of miR-34a contributes to p53-mediated apoptosis. Mol Cell 26:731–743
42. Yang CH, Yue J, Fan M, Pfeffer LM (2010) Interferon induces miR-21 through a STAT3-dependent pathway as a suppressive negative feedback on interferon-induced apoptosis. Cancer Res 70(20):8108–8116. Epub 2010 Sep 2
43. Loven J, Zinin N, Wahlstrom T, Muller I, Brodin P, Fredlund E, Ribacke U, Pivarcsi A, Pahlman S, Henriksson M (2010) MYCN-regulated microRNAs repress estrogen receptor-alpha (ESR1) expression and neuronal differentiation in human neuroblastoma. Proc Natl Acad Sci U S A 107:1553–1558
44. Yan HL, Xue G, Mei Q, Wang YZ, Ding FX, Liu MF, Lu MH, Tang Y, Yu HY, Sun SH (2009) Repression of the miR-17-92 cluster by p53 has an important function in hypoxia-induced apoptosis. Embo J 28:2719–2732

45. Shan SW, Lee DY, Deng Z, Shatseva T, Jeyapalan Z, Du WW, Zhang Y, Xuan JW, Yee SP, Siragam V, Yang BB (2009) MicroRNA MiR-17 retards tissue growth and represses fibronectin expression. Nat Cell Biol 11:1031–1038
46. Petrocca F, Vecchione A, Croce CM (2008) Emerging role of miR-106b-25/miR-17-92 clusters in the control of transforming growth factor beta signaling. Cancer Res 68:8191–8194
47. Liu G, Friggeri A, Yang Y, Milosevic J, Ding, Q, Thannickal, VJ, Kaminski N, Abraham E (2010) miR-21 mediates fibrogenic activation of pulmonary fibroblasts and lung fibrosis. J Exp Med 207:1589–1597
48. Lu J, Getz G, Miska EA, Alvarez-Saavedra E, Lamb J, Peck D, Sweet-Cordero A, Ebert BL, Mak RH, Ferrando AA, Downing JR, Jacks T, Horvitz HR, Golub TR (2005) MicroRNA expression profiles classify human cancers. Nature 435:834–838
49. Ma L, Teruya-Feldstein J, Weinberg RA (2007) Tumour invasion and metastasis initiated by microRNA-10b in breast cancer. Nature 449:682–688
50. Sachdeva M, Mo YY (2010) MicroRNA-145 suppresses cell invasion and metastasis by directly targeting mucin 1. Cancer Res 70:378–387
51. Hwang HW, Mendell JT (2006) MicroRNAs in cell proliferation, cell death, and tumorigenesis. Br J Cancer 94:776–780
52. Hammond SM (2006) MicroRNAs as oncogenes. Curr Opin Genet Dev 16:4–9
53. Gregory RI, Shiekhattar R (2005) MicroRNA biogenesis and cancer. Cancer Res 65:3509–3512
54. Schetter AJ, Leung SY, Sohn JJ, Zanetti KA, Bowman ED, Yanaihara N, Yuen ST, Chan TL, Kwong DL, Au GK, Liu CG, Calin GA, Croce CM, Harris CC (2008) MicroRNA expression profiles associated with prognosis and therapeutic outcome in colon adenocarcinoma. Jama 299:425–436
55. Jiang J, Gusev Y, Aderca I, Mettler TA, Nagorney DM, Brackett DJ, Roberts LR, Schmittgen TD (2008) Association of MicroRNA expression in hepatocellular carcinomas with hepatitis infection, cirrhosis, and patient survival. Clin Cancer Res 14:419–427
56. Mathe EA, Nguyen GH, Bowman ED, Zhao Y, Budhu A, Schetter AJ, Braun R, Reimers M, Kumamoto K, Hughes D, Altorki NK, Casson AG, Liu CG, Wang XW, Yanaihara N, Hagiwara N, Dannenberg AJ, Miyashita M, Croce CM, Harris CC (2009) MicroRNA expression in squamous cell carcinoma and adenocarcinoma of the esophagus: associations with survival. Clin Cancer Res 15:6192–6200
57. Hui AB, Lenarduzzi M, Krushel T, Waldron L, Pintilie M, Shi W, Perez-Ordonez B, Jurisica I, O'Sullivan B, Waldron J, Gullane P, Cummings B, Liu FF (2010) Comprehensive MicroRNA profiling for head and neck squamous cell carcinomas. Clin Cancer Res 16:1129–1139
58. Selaru FM, Olaru AV, Kan T, David S, Cheng Y, Mori Y, Yang J, Paun B, Jin Z, Agarwal R, Hamilton JP, Abraham J, Georgiades C, Alvarez H, Vivekanandan P, Yu W, Maitra A, Torbenson M, Thuluvath PJ, Gores GJ, LaRusso NF, Hruban R, Meltzer SJ (2009) MicroRNA-21 is overexpressed in human cholangiocarcinoma and regulates programmed cell death 4 and tissue inhibitor of metalloproteinase 3. Hepatology 49:1595–1601
59. Chan JA, Krichevsky AM, Kosik KS (2005) MicroRNA-21 is an antiapoptotic factor in human glioblastoma cells. Cancer Res 65:6029–6033
60. Si ML, Zhu S, Wu H, Lu Z, Wu F, Mo YY (2007) miR-21-mediated tumor growth. Oncogene 26:2799–2803
61. Cottonham CL, Kaneko S, Xu L (2010) miR-21 and miR-31 converge on TIAM1 to regulate migration and invasion of colon carcinoma cells. J Biol Chem 285(46):35293–35302. Epub 2010 Sep 7 2010
62. Sarkar J, Gou D, Turaka P, Viktorova E, Ramchandran R, Raj JU (2010) MicroRNA-21 plays a Role in Hypoxia-mediated Pulmonary Artery Smooth Muscle Cell Proliferation and Migration. Am J Physiol Lung Cell Mol Physiol 299(6):L861–871. Epub 2010 Aug 6
63. Meng F, Henson R, Wehbe-Janek H, Ghoshal K, Jacob ST, Patel T (2007) MicroRNA-21 regulates expression of the PTEN tumor suppressor gene in human hepatocellular cancer. Gastroenterology 133:647–658

64. Zhu S, Si ML, Wu H, Mo YY (2007) MicroRNA-21 targets the tumor suppressor gene tropomyosin 1 (TPM1). J Biol Chem 282:14328–14336
65. Frankel LB, Christoffersen NR, Jacobsen A, Lindow M, Krogh A, Lund AH (2008) Programmed cell death 4 (PDCD4) is an important functional target of the microRNA miR-21 in breast cancer cells. J Biol Chem 283:1026–1033
66. Zhu S, Wu H, Wu F, Nie D, Sheng S, Mo YY (2008) MicroRNA-21 targets tumor suppressor genes in invasion and metastasis. Cell Res 18:350–359
67. Asangani IA, Rasheed SA, Nikolova DA, Leupold JH, Colburn NH, Post S, Allgayer H (2008) MicroRNA-21 (miR-21) post-transcriptionally downregulates tumor suppressor Pdcd4 and stimulates invasion, intravasation and metastasis in colorectal cancer. Oncogene 27:2128–2136
68. Lu Z, Liu M, Stribinskis V, Klinge CM, Ramos KS, Colburn NH, Li Y (2008) MicroRNA-21 promotes cell transformation by targeting the programmed cell death 4 gene. Oncogene 27:4373–4379
69. Hatley ME, Patrick DM, Garcia MR, Richardson JA, Bassel-Duby R, van Rooij E, Olson EN (2010) Modulation of K-Ras-dependent lung tumorigenesis by MicroRNA-21. Cancer Cell 18:282–293
70. Medina PP, Nolde M, Slack FJ (2010) OncomiR addiction in an *in vivo* model of microRNA-21-induced pre-B-cell lymphoma. Nature 467:86–90
71. He L, Thomson JM, Hemann MT, Hernando-Monge E, Mu D, Goodson S, Powers S, Cordon-Cardo C, Lowe SW, Hannon GJ, Hammond SM (2005) A microRNA polycistron as a potential human oncogene. Nature 435:828–833
72. Olive V, Bennett MJ, Walker JC, Ma C, Jiang I, Cordon-Cardo C, Li QJ, Lowe SW, Hannon GJ, He L (2009) miR-19 is a key oncogenic component of mir-17–92. Genes Dev 23:2839–2849
73. Mu P, Han YC, Betel D, Yao E, Squatrito M, Ogrodowski P, de Stanchina E, D'Andrea A, Sander C, Ventura A (2009) Genetic dissection of the miR-17~92 cluster of microRNAs in Myc-induced B-cell lymphomas. Genes Dev 23:2806–2811
74. Mavrakis KJ, Wolfe AL, Oricchio E, Palomero T, de Keersmaecker K, McJunkin K, Zuber J, James T, Khan AA, Leslie CS, Parker JS, Paddison PJ, Tam W, Ferrando A, Wendel HG (2010) Genome-wide RNA-mediated interference screen identifies miR-19 targets in Notch-induced T-cell acute lymphoblastic leukaemia. Nat Cell Biol 12:372–379
75. Mi S, Li Z, Chen P, He C, Cao D, Elkahloun A, Lu J, Pelloso LA, Wunderlich M, Huang H, Luo RT, Sun M, He M, Neilly MB, Zeleznik-Le NJ, Thirman MJ, Mulloy JC, Liu PP, Rowley JD, Chen J (2010) Aberrant overexpression and function of the miR-17–92 cluster in MLL-rearranged acute leukemia. Proc Natl Acad Sci U S A 107:3710–3715
76. Hayashita Y, Osada H, Tatematsu Y, Yamada H, Yanagisawa K, Tomida S, Yatabe Y, Kawahara K, Sekido Y, Takahashi TA (2005) polycistronic microRNA cluster, miR-17-92, is overexpressed in human lung cancers and enhances cell proliferation. Cancer Res 65:9628–9632
77. Zhang X, Ladd A, Dragoescu E, Budd WT, Ware JL, Zehner ZE (2009) MicroRNA-17-3p is a prostate tumor suppressor *in vitro* and *in vivo*, and is decreased in high grade prostate tumors analyzed by laser capture microdissection. Clin Exp Metastasis 26:965–979
78. Yang F, Yin Y, Wang F, Wang Y, Zhang L, Tang Y, Sun S (2010) miR-17-5p Promotes migration of human hepatocellular carcinoma cells through the p38 mitogen-activated protein kinase-heat shock protein 27 pathway. Hepatology 51:1614–1623
79. Chow TF, Mankaruos M, Scorilas A, Youssef Y, Girgis A, Mossad S, Metias S, Rofael Y, Honey RJ, Stewart R, Pace KT, Yousef GM (2010) The miR-17-92 cluster is over expressed in and has an oncogenic effect on renal cell carcinoma. J Urol 183:743–751
80. Bloomston M, Frankel W L, Petrocca F, Volinia S, Alder H, Hagan JP, Liu CG, Bhatt D, Taccioli C, Croce CM (2007) MicroRNA expression patterns to differentiate pancreatic adenocarcinoma from normal pancreas and chronic pancreatitis. Jama 297:1901–1908
81. Yan LX, Huang XF, Shao Q, Huang MY, Deng L, Wu QL, Zeng YX, Shao JY (2008) MicroRNA miR-21 overexpression in human breast cancer is associated with advanced clinical stage, lymph node metastasis and patient poor prognosis. Rna 14:2348–2360

82. Huang TH, Wu F, Loeb GB, Hsu R, Heidersbach A, Brincat A, Horiuchi D, Lebbink RJ, Mo YY, Goga A, McManus MT (2009) Up-regulation of miR-21 by HER2/neu signaling promotes cell invasion. J Biol Chem 284:18515–18524
83. Baffa R, Fassan M, Volinia S, O'Hara B, Liu CG, Palazzo JP, Gardiman M, Rugge M, Gomella LG, Croce CM, Rosenberg A (2009) MicroRNA expression profiling of human metastatic cancers identifies cancer gene targets. J Pathol 219:214–221
84. Gabriely G, Wurdinger T, Kesari S, Esau CC, Burchard J, Linsley PS, Krichevsky AM (2008) MicroRNA 21 promotes glioma invasion by targeting matrix metalloproteinase regulators. Mol Cell Biol 28:5369–5380
85. Reis PP, Tomenson M, Cervigne NK, Machado J, Jurisica I, Pintilie M, Sukhai MA, Perez-Ordonez B, Grenman R, Gilbert RW, Gullane PJ, Irish JC, Kamel-Reid S (2010) Programmed cell death 4 loss increases tumor cell invasion and is regulated by miR-21 in oral squamous cell carcinoma. Mol Cancer 9:238
86. Yu Z, Willmarth NE, Zhou J, Katiyar S, Wang M, Liu Y, McCue PA, Quong AA, Lisanti MP, Pestell RG (2010) microRNA 17/20 inhibits cellular invasion and tumor metastasis in breast cancer by heterotypic signaling. Proc Natl Acad Sci U S A 107:8231–8236
87. Yu J, Ohuchida K, Mizumoto K, Fujita H, Nakata K, Tanaka M (2010) MicroRNA miR-17-5p is overexpressed in pancreatic cancer, associated with a poor prognosis and involved in cancer cell proliferation and invasion. Cancer Biol Ther 10(8):748–757. Epub 2010 Oct 15
88. Li H, Bian C, Liao L, Li J, Zhao RC (2011) miR-17-5p promotes human breast cancer cell migration and invasion through suppression of HBP1. Breast Cancer Res Treat 126(3):565–575. Epub 2010 May 27
89. Liu S, Goldstein RH, Scepansky EM, Rosenblatt M (2009) Inhibition of rho-associated kinase signaling prevents breast cancer metastasis to human bone. Cancer Res 69:8742–8751
90. Sachdeva M, Wu H, Ru P, Hwang L, Trieu V, Mo YY MicroRNA-101-mediated Akt activation and estrogen-independent growth. Oncogene 2010 [Epub ahead of print]
91. Wang P, Zou F, Zhang X, Li H, Dulak A, Tomko RJ, Jr Lazo JS, Wang Z, Zhang L, Yu J (2009) microRNA-21 negatively regulates Cdc25A and cell cycle progression in colon cancer cells. Cancer Res 69:8157–8165
92. Hossain A, Kuo MT, Saunders GF (2006) Mir-17-5p regulates breast cancer cell proliferation by inhibiting translation of AIB1 mRNA. Mol Cell Biol 26:8191–8201
93. Cloonan N, Brown MK, Steptoe AL, Wani S, Chan WL, Forrest AR, Kolle G, Gabrielli B, Grimmond SM (2008) The miR-17-5p microRNA is a key regulator of the G1/S phase cell cycle transition. Genome Biol 9:R127
94. Meng F, Henson R, Lang M, Wehbe H, Maheshwari S, Mendell JT, Jiang J, Schmittgen TD, Patel T (2006) Involvement of human micro-RNA in growth and response to chemotherapy in human cholangiocarcinoma cell lines. Gastroenterology 130:2113–2129
95. Hwang JH, Voortman J, Giovannetti E, Steinberg SM, Leon LG, Kim YT, Funel N, Park JK, Kim MA, Kang GH, Kim SW, Del Chiaro M, Peters GJ, Giaccone G (2010) Identification of microRNA-21 as a biomarker for chemoresistance and clinical outcome following adjuvant therapy in resectable pancreatic cancer. PLoS One 5:e10630
96. Li Y, Zhu X, Gu J, Dong D, Yao J, Lin C, Huang K, Fei J (2010) Anti-miR-21 oligonucleotide sensitizes leukemic K562 cells to arsenic trioxide by inducing apoptosis. Cancer Sci 101:948–954
97. Li Y, Zhu X, Gu J, Hu H, Dong D, Yao J, Lin C, Fei J (2010) Anti-miR-21 oligonucleotide enhances chemosensitivity of leukemic HL60 cells to arabinosylcytosine by inducing apoptosis. Hematology 15:215–221
98. Shi L, Chen J, Yang J, Pan T, Zhang S, Wang Z (2010) MiR-21 protected human glioblastoma U87MG cells from chemotherapeutic drug temozolomide induced apoptosis by decreasing Bax/Bcl-2 ratio and caspase-3 activity. Brain Res 1352:255–264
99. Zhang HL, Yang LF, Zhu Y, Yao XD, Zhang SL, Dai B, Zhu YP, Shen YJ, Shi GH, Ye DW (2011) Serum miRNA-21:Elevated levels in patients with metastatic hormone-refractory prostate cancer and potential predictive factor for the efficacy of docetaxel-based chemotherapy. Prostate 71(3):326–331. doi: 10.1002/pros.21246. Epub 2010 Sep 14

100. Ferracin M, Zagatti B, Rizzotto L, Cavazzini F, Veronese A, Ciccone M, Saccenti E, Lupini L, Grilli A, De Angeli C, Negrini M, Cuneo A (2010) MicroRNAs involvement in fludarabine refractory chronic lymphocytic leukemia. Mol Cancer 9:123
101. Fontana L, Fiori ME, Albini S, Cifaldi L, Giovinazzi S, Forloni M, Boldrini R, Donfrancesco A, Federici V, Giacomini P, Peschle C, Fruci D (2008) Antagomir-17-5p abolishes the growth of therapy-resistant neuroblastoma through p21 and BIM. PLoS One 3:e2236
102. Slaby O, Svoboda M, Fabian P, Smerdova T, Knoflickova D, Bednarikova M, Nenutil R, Vyzula R (2007) Altered expression of miR-21, miR-31, miR-143 and miR-145 is related to clinicopathologic features of colorectal cancer. Oncology 72:397–402
103. Yantiss RK, Goodarzi M, Zhou XK, Rennert H, Pirog EC, Banner BF, Chen YT (2009) Clinical, pathologic, and molecular features of early-onset colorectal carcinoma. Am J Surg Pathol 33(4):572–582
104. Markou A, Tsaroucha E G, Kaklamanis L, Fotinou M, Georgoulias V, Lianidou ES (2008) Prognostic value of mature microRNA-21 and microRNA-205 overexpression in non-small cell lung cancer by quantitative real-time RT-PCR. Clin Chem 54:1696–1704
105. Mitchell PS, Parkin RK, Kroh EM, Fritz BR, Wyman SK, Pogosova-Agadjanyan EL, Peterson A, Noteboom J, O'Briant KC, Allen A, Lin DW, Urban N, Drescher CW, Knudsen BS, Stirewalt DL, Gentleman R, Vessella RL, Nelson PS, Martin DB, Tewari M (2008) Circulating microRNAs as stable blood-based markers for cancer detection. Proc Natl Acad Sci U S A 105:10513–10518
106. Davis S, Lollo B, Freier S, Esau C (2006) Improved targeting of miRNA with antisense oligonucleotides. Nucleic Acids Res 34:2294–2304
107. Elmen J, Lindow M, Schutz S, Lawrence M, Petri A, Obad S, Lindholm M, Hedtjarn M, Hansen HF, Berger U, Gullans S, Kearney P, Sarnow P, Straarup EM, Kauppinen S (2008) LNA-mediated microRNA silencing in non-human primates. Nature 452:896–899
108. Bandyopadhyay S, Mitra R, Maulik U, Zhang MQ (2010) Development of the human cancer microRNA network. Silence 1:6
109. Asangani IA, Rasheed SA, Nikolova DA, Leupold JH, Colburn NH, Post S, Allgayer H (2008) MicroRNA-21 (miR-21) post-transcriptionally downregulates tumor suppressor Pdcd4 and stimulates invasion, intravasation and metastasis in colorectal cancer. Oncogene. 27(15):2128–2136. Epub 2007 Oct 29
110. Klusmann JH, Li Z, Bohmer K, Maroz A, Koch ML, Emmrich S, Godinho FJ, Orkin SH, Reinhardt D (2010) miR-125b-2 is a potential oncomiR on human chromosome 21 in megakaryoblastic leukemia. Genes Dev 24:478–490
111. Kong W, He L, Coppola M, Guo J, Esposito NN, Coppola D, Cheng JQ (2010) MicroRNA-155 regulates cell survival, growth, and chemosensitivity by targeting FOXO3a in breast cancer. J Biol Chem 285:17869–17879
112. Galardi S, Mercatelli N, Giorda E, Massalini S, Frajese GV, Ciafre SA, Farace MG (2007) miR-221 and miR-222 expression affects the proliferation potential of human prostate carcinoma cell lines by targeting p27Kip1. J Biol Chem 282:23716–23724
113. Huang Q, Gumireddy K, Schrier M, le Sage C, Nagel R, Nair S, Egan DA, Li A, Huang G, Klein-Szanto AJ, Gimotty PA, Katsaros D, Coukos G, Zhang L, Pure E, Agami R (2008) The microRNAs miR-373 and miR-520c promote tumour invasion and metastasis. Nat Cell Biol 10:202–210
114. de Souza Rocha Simonini P, Breiling A, Gupta N, Malekpour M, Youns M, Omranipour R, Malekpour F, Volinia S, Croce CM, Najmabadi H, Diederichs S, Sahin O, Mayer D, Lyko F, Hoheisel JD, Riazalhosseini Y (2010) Epigenetically Deregulated microRNA-375 Is Involved in a Positive Feedback Loop with Estrogen Receptor {alpha} in Breast Cancer Cells. Cancer 70(22):9175–9184. Epub 2010 Oct 26
115. Hu W, Chan CS, Wu R, Zhang C, Sun Y, Song JS, Tang LH, Levine AJ, Feng Z (2010) Negative regulation of tumor suppressor p53 by microRNA miR-504. Mol Cell 38:689–699

Regulation of Metastasis by miRNAs

Suresh K. Alahari

Abstract A breakthrough of the twenty-first century is the discovery in which protein expression is regulated by a new class of endogenous, noncoding, small RNAs (microRNAs). These microRNAs (miRNAs) control gene expression by acting on their target mRNAs, inducing either mRNA degradation or translational repression. MiRNAs are single-stranded and highly conserved between species. MiRNAs have been implicated in the detection and treatment of various pathological states. In the genome, most miRNAs are much smaller than protein coding genes, but they have specific functions in various biological processes. The major process in cancer progression is metastasis. Metastatic processes include invasion, intravasation, and extravasation. In this chapter, we introduce basic information about miRNAs, their functional mechanism, and their biological roles in tumor metastasis.

1 MiRNAs: The Basics

1.1 Discovery

In 1993, Ambros et al. first discovered miRNAs [1]; in a genetic screening it was found that round worm *Caenorhabditis elegans*, with one gene, lin-4, did not encode a protein, but instead a 22-nucleotide small RNA. In 2000, another miRNA (highly conserved between species) with 22-nucleotide, let-7, was discovered in *C. elegans*. It has been suggested that let 7 has a major biological function. MiRNA Let-7 has been reported to be involved in developmental coordination [2, 3]. Most of the multicellular organisms were subsequently discovered to express regulatory miRNAs similar to lin-4 and let-7 [4, 5, 6]. Involvement of miRNAs in cancer was first discovered by Croce et al. [7], who discovered miR-15 and miR-16. These two miRNAs are frequently deleted in human chronic lymphocytic leukemias. MiRNAs

S. K. Alahari (✉)
Department of Biochemistry and Molecular Biology, Stanley S. Scott Cancer Center,
LSU School of Medicine, New Orleans, LA 70112, USA
e-mail: salaha@lsuhsc.edu

S. Alahari (ed.), *MicroRNA in Cancer*,
DOI 10.1007/978-94-007-4655-8_6,

are also found as novel biomarkers of cancer, including breast cancer [8]. Jean S. Kan et al. demonstrated that 28 miRNAs are involved in modulation of invasion in various cancers [9]. Ming Shi et al. reported that alteration in the expression of miRNAs leads to cancer pathogenesis (including that of breast cancer) through modification of differentiation, cell proliferation, apoptosis, and metastasis [10]. MiRNAs that regulate cancer can be divided into two groups, oncogenic miRNAs and tumor suppressor miRNAs [11]. These reports indicate that miRNAs have pivotal role in several biological processes.

1.2 Biogenesis

In the nucleus, RNA polymerase II makes primary transcripts (500–3,000 bases) of miRNAs (pri-miRNA) from their genes. Subsequently, pre-miRNAs (~70 bases) are derived from pri-miRNAs in a process catalyzed by endonucleases of the RNAse III family, Drosha in flies and DGCR8 in humans. Pre-miRNAs formed in the nucleus are exported by exportin and Ran-GTP, after which another RNAse III, Dicer, acts on miRNA to generate ~22 nucleotide double-stranded miRNA duplexes. Each of the duplexes forms a complex with RNA-induced silencing complex (RISC) to unwind mature miRNA (single-stranded miRNA). So far, more than 1,000 miRNAs have been found in humans (miR Base Sequence Database-Release 16). About 5,300 human genes have been predicted as miRNA targets, representing 30 % of the human gene set.

1.3 Mechanism of miRNA Gene Regulation

At least one-third of human genes believed to be regulated by miRNAs. Expression of target genes of miRNAs is down-regulated by many mature miRNAs. Based on the complementary sequence, miRNAs recognize their target mRNAs. The miRNA-RISC complex binds with mRNA and subsequently inhibits protein translation and/or degradation of mRNA to modify cellular response. Mechanistically, miRNAs regulate gene expression either by translation inhibition or mRNA destabilization. MiRNA binds either with 3′UTR (untranslated region) or the open reading frame (ORF) of target mRNA; the status of binding depends on complementary sequences between them. Imperfect complementary sequences with 3′UTR cause translational repression. Binding with ORF is mostly by perfect complementary sequences to cause either cleavage or degradation of mRNA with Argonaute2. Recent evidence explains other mechanisms that are involved in these processes, such as mRNA degradation, which occurs only through partial complementary sequences between miRNA and target mRNA [12].

miRNAs have been reported to interfere in cell signaling events with TGF-β, WNT, Notch, and EGF [13]. Cell-cycle regulators are directly regulated by miR-15a/16 cluster, miR-17/20 cluster, miR-221/222 cluster, let-7, and miR-34 families to regulate cell cycle progression [14]. In several tumor types, a group of miRNAs that are encoded by the miR-17~92 cluster are implicated as oncogenes. Uziel et al. have reported that there is a functional collaboration between the miR-17~92 cluster and sonic hedgehog signaling in medulloblastoma development [15]. MiR-34 was found to be a transcriptional target of p53, and its activity leads to induction of cell cycle arrest and promotion of apoptosis. Loss of miR-34 expression impairs p53-mediated cell death [16]. Aggressive types of nonsmall cell lung cancer and hepatocellular carcinomas overexpress miR-221/222 compared to that of less invasive and/or normal cells. The miRNAs act on their targets, PTEN, TIMP3; they also induce TNF-related apoptosis-inducing ligand (TRAIL) resistance and activate AKT signaling, as well as metallopeptidases to enhance cellular migration [17]. It also has been shown that each miRNA can regulate various mRNAs and that each mRNA can be targeted by various miRNAs [18]. MiRNAs differentially regulate the proliferation and death of normal and cancer cells and therefore are considered to be major contributors to cancer pathogenesis [19]. White et al. stated that miRNAs are involved in potential mechanisms for tumor aggressiveness such as the transition of epithelial cells to mesenchymal cells, angiogenesis, and apoptosis [20]. MiR-200 family members are also involved in regulating the epithelial-to-mesenchymal transition [21].

2 MicroRNAs and Metastasis

In most cancer patients, recurrence and metastasis in most of the malignancies is associated with death. The multistep process of metastasis includes the invasion of malignant cells into the microenvironment, their entry into the bloodstream, migration, and extravasation into distant organs. In addition, metastasis is followed by subsequent steps of proliferation, angiogenesis, and evasion of apoptosis; these are crucial events for the colonization of a secondary site. Therefore, metastatic events comprise various genetic and epigenetic changes. Baranwal and Alahari reported that miRNAs regulate invasion, migration, and metastasis in cancers. Dysregulation of miRNAs promotes cell cycle progression, prevents apoptosis of cancer cells, and induces metastasis [22]. Such abnormal miRNA regulation was found in gastric cancer tissues [23]. MiRNA genes of the miR-200 family are frequently downregulated in tumor progression. These genes are involved in differential regulation of EGF-driven invasion, viability, apoptosis, and cell cycle progression in breast cancer cells to suppress tumor development [24]. In many breast cancers, estrogen receptor β1 (ERβ1) is down-regulated. In a mouse model, expression of miR-193b significantly inhibits xenograft tumors and dissemination of tumors. MiR-193b is a negative regulator of the uPA gene (target gene) in primary breast tumors [25]. In many metastatic cancers, including oral squamous-cell carcinoma (OSCC), tumor-suppressor programmed cell death 4 (PDCD4) was found to be under- expressed, with miR-21 mediating the underexpresson [26]. The miR-196 family includes miR-196a1, miR-196a2, and miR-196b; these miRNAs have been shown to reduce invasion and metastasis in breast cancer models [27]. A p53 family member, Tap63 suppresses tumor progression and metastasis; it also coordinately regulates Dicer and miR-130b. Modification in expression of Dicer affects metastatic potential through abnormal coordination with miR-130b in tumorigenesis [28]. Two miRNAs let-7 and miR-22 were reported to be downregulated in metastatic breast cancers; their target genes ERBB3, CDC25C and EVI-1 were found to be upregulated in cancers [29].

MiR-22 is highly expressed in human epithelial cells and senescent fibroblasts, but in many cancer cell lines it is downregulated. Overexpression of miR-22 leads to suppression of growth; its knockdown in presenescent fibroblasts reduces cell size and motility, and inhibits invasion. It has been confirmed that miR-22 acts as a tumor suppressor by inducing senescence; it does this by targeting CDK6, SIRT1, and Sp1 [30]. MiR-194 has high expression in the human gastrointestinal tract and kidney. Overexpression of miR-194 in liver mesenchymal-like cancer cells leads to reduction of the mesenchymal marker N-cadherin, as well as the suppression of invasion and migration of cancer cells [31]. MiR-520h mediates cancer cell migration; it is downregulated by the adenovirus type 5 E1A, which has been used in clinical trials to treat cancer. This miRNA induces a tumorigenic pathway through activating twist, a protein that regulates E-cadherin, an epithelial marker [32].

A microarray analysis with 43 paired primary tumors showed that the following miRNAs are associated with cancer progression: miR-10b, miR-21, miR-30a, miR-30e, miR-125b, miR-141, miR-200b, miR-200c and miR-205 [33]. Generally, miR-200 suppresses invasion and metastasis in cancer [34]. MiR-145 functions as a tumor suppressor, inhibiting tumor cell growth *in vitro* and *in vivo*. In animal models, miR-145 suppresses lung metastasis by partially suppressing its target gene mucin1 (a metastasis gene). The suppression of mucin1 leads to reduction of β-catenin and oncogenic cadherin11 [35]. The miR-17/20 cluster regulates migration and invasion through heterotypic secreted signals. MiR-17/20 was found to be downregulated in breast cancer cells. These miRNAs repress IL-8 and inhibit cytokeratin 8 through cyclin D1 to regulate tumor metastasis [36]. As discussed above, miRNAs regulate gene expression by targeting mRNAs. Some miRNAs are more highly expressed in cancer tissues than in their respective normal tissues. MiR-125b, miR-145, miR-21 and miR-155 are found to be highly expressed (deregulated) in cancer tissues such as breast cancers [37].

2.1 Function of Specific miRNAs in EMT

One important requirement for cancer metastasis is epithelial mesenchymal transition (EMT), a cellular process that converts immotile, polarized epithelial cells into motile mesenchymal cells. One of the cell adhesion proteins, E-cadherin, is a major constituent of adherens junctions and seems to function as a suppressor of invasion or metastasis during carcinoma development. Recent reports indicate that members of the miR-200 family (miR-141, miR-200b, and miR-200c) are key mediators in regulatory events in the expression of E-cadherin [38]. MiR-9 downregulates E-cadherin to activate β-catenin signaling, which leads to the induction of VEGF upregulation and tumor angiogenesis [39]. MiR-132 has been found to be a switch for the angiogenic process, inducing neovascularization by targeting p120RasGAP in the endothelium in order to induce Ras function [40]. Mesenchymal, metastatic RasXT cells upregulate miR-29a, which suppresses a protein, tristetraprolin (TTP), which is involved in mRNA degradation. Enhanced miR-29a expression reduces TTP in breast cancer. Moreover, miR-29a disturbs epithelial polarity to induce metastasis [41]. Many transcription factors of the zinc finger family, such as Snail1, Slug, ZEB1, and ZEB2, have been shown to mediate the process of epithelial-mesenchymal transition. The regulatory pattern of the miR-200 family on these zinc finger transcription factors has been demonstrated. Highly invasive mesenchymal cells express ZEB family transcription factors. Expression of ZEBs is directly repressed by the miR-200 family (miR-141, miR-200b and miR-200c) in noninvasive epithelial cells. ZEBs repress the transcription of miR-200 genes; here it is clear that there is formation of a double-negative feedback loop to ensure the repression of epithelial genes in ZEBs-expressing mesenchymal cells. This suggests that miR-200 is a major regulator of EMT [42, 43, 44]. Therefore, modulation of E-cadherin expression is an attractive idea for cancer therapies, and both ZEBs and miR-200 are potentially good targets for further study.

The cell-cell adhesion protein Nectin-1 and lipid transferase StarD10 are targets of miR-661. These two targets are involved in invasion processes in breast cancer. In particular, the expression of StarD10 is associated with markers of luminal subtypes of breast cancer cells. MiR-661 downregulates these two targets to contribute to invasion of breast cancer cells [45]. Another miRNA, miR-101, is also involved in EMT, through the enhancer of zeste homolog (EZH2). EZH2 is a histone methyl transferase that contributes to the process of epigenetic silencing of E-cadherin and other target genes. It regulates survival and metastasis of cancer cells [46]. It is known that EZH2 is repressed by miRNA-101; most aggressive and clinically localized prostate cancers show high expression of miRNA-101. It has also been found that the locus of miRNA-101 is lost in metastatic prostate cancers. In turn, upregulation of EZH2 and misregulation of epigenetic pathways leads to cancer progression [47]. Epigenetic pathways can be regulated by miRNA-101. MiR-155 has been reported to express differentially among normal murine mammary gland (NMuMG) epithelial cells and TGFβ-treated NMuMG cells. In advanced malignancies, miR-155 expression and its promoter activity are believed to be induced by TGF-β through Smad4. MiR-155 expression has been detected in invasive breast cancer tissues. TGF-β mediated EMT migration and invasion are suppressed by knockdown of miR-155; ectopic expression of miR-155 disrupts tight junction formation and leads to diminished RhoA expression. Therefore, the TGF-β/Smad pathway is involved in regulation of miR-155 function.

2.2 miRNAs Drive the Cancer Metastasis

In cancer progression, cell invasion is the first step. As noted earlier, cancer development involves the movement of tumor cells from their place of origin to neighboring tissues. Tumor cell migration and dissolution of extracellular matrix proteins play an important role in invasion, and many miRNAs are involved in the processes of metastasis and some important ones are discussed below.

2.2.1 miR-21

In many cancers, miR-21 is involved in cell invasion and migration. miR is important for carcinogenesis and metastasis [48]. Studies on colorectal and breast cancers have demonstrated that miR-21 posttranscriptionally down-regulates tumor suppressor programmed cell death 4 (PDCD4), whereas tropomyosin 1 (TPM1) genes stimulate invasion and metastasis. In addition, miR-21 is negatively correlated with expression of bone morphogenetic protein-6 (BMP-6) in breast cancer tissue samples. BMP-6 regulates miR-21 expression in MDA-MB-231 cells at the transcriptional level. This is mediated by repression of δEF1 and c-Fos/c-Jun signaling. BMP-6 binds with miR-21 promoter to inhibit δEF1 and c-Fos/c- Jun [49]. MiR-21 expression is also correlated with upregulation of HER2/neu. Thus, in breast cancer, upon stimulation of Her2/neu, miR-21 is functionally involved in HER2/neu-mediated cell invasion (mediated through the MAPK (ERK1/2) pathway) [50].

It has been found in prostate cancer cell lines (DU145 and PC-3) that cell adhesion and motility are also regulated by miR-21 through the targeting of myristoylated alanine-rich protein kinase C substrate (MARCKS) in order to modify the actin skeleton [51]. In hepatocellular cancer, miR-21 upregulates tumor cell proliferation, migration, and invasion, but downregulates tumor suppressor expression of the phosphatase and tensin homologs (PTEN). Meng and colleagues (2007) have shown that modulation of miR-21 activity leads to alteration in focal adhesion kinase (FAK) phosphorylation and expression of matrix metalloproteases 2 and 9. These two metalloproteases are downstream mediators of PTEN; they are involved in processes for tumor cell migration and invasion [52]. MiR-21 has been found to have high expression in nonsmall cell lung cancer. It is also a marker for poor prognosis. MiR-21 mediates tumorigenesis by inhibiting apoptosis and negative regulators of Ras/MEK/ERK signaling [53].

In glioma cells, RECK and TIMP3 genes are malignancy suppressors and inhibitors for matrix metalloproteinases. These two genes are regulated by miR-21. Repression of miR-21 in glioma cells leads to a reduction in the invasion and migration of the cells, and this explores that miR-21 induces glioma malignancy by down- regulating MMP inhibitors like RECK and TIMP3 [54]. In summary, because miR-21 has multiple targets in various types of cancers, it is an important regulator of various genes in cancer.

2.2.2 miR-10b

In breast cancer metastasis, miR-10b was found to be a critical molecule [55]. In vitro experiments showed that absence of miR-10b reduces invasion significantly. MiR-10b ectopic expression does not affect cell proliferation, but increases invasion and migration of immortalized breast cells HMECs and SUM149. MiR-10b does not affect primary tumor growth. MiR-10b-overexpressing tumors exhibit invasion fronts having very high cell proliferative and angiogenic activities. The EMT promoter Twist1 has been reported to have a direct role in the expression of miR-10b regulation; Twist1 overexpression augments expression of miR-10b. In the case of another EMT regulator, snail1, its overexpression has no effect on miR-10b. Two other functional targets of miR-10b are homeoboxD10 (HOXD10) and RB1CC1 (also called FIP200). These two targets are implicated in the suppression of cell migration and/or invasion. In order to reduce migration and/or invasion, HOXD10 downregulates the expression of genes involved in migration or invasion such as RHOC, α3 integrin, matrix metalloproteinase-14, and urokinase-type plasminogen activator receptor. MiR-10b is specifically involved in the metastatic process, but not in primary tumor formation. Thus, it is possible to develop miRNA-based drugs that are specific against metastasis [55].

2.2.3 miR-182, miR-183, miR-380-5p and miR-211

One highly aggressive form of cancers is melanoma; its mechanism of development is poorly understood. MicroRNA profiling of normal melanocytes and melanoma cell lines indicate that miR-182, a prometastatic miRNA, is a critically important miRNA. It is flanked by the oncogenes c-MET and BRAF, and is located in the 7q31–34 region, which is frequently amplified in melanomas. In melanoma, down-regulation of miR-182 reduces invasion and triggers apoptosis; its overexpression promotes migration and survival by repressing microphthalmia-associated transcription factor-M and FOXO3. Overexpression of either microphthalmia-associated transcription factor-M or FOXO3 blocks the invasive effects of miR-182 [56, 57]. Therefore, microphthalmia-associated transcription factor-M and FOXO3 are potential targets of miR-182. This provides good therapeutic strategy for melanoma through miR-182.

MiR-183 suppresses the expression of ezrin in lung cancer. Ezrin is a member of the ERM (Ezrin, Radixin, and Moesin) group of proteins; these have been identified as critical regulators of cell migration and metastasis. MiR-183 is located on chromosome 7q32 and is part of the miRNA family, which is abnormally regulated in many cancers [58]. VIL2, another target of miR-183, may also be involved in the regulation of migration and metastasis in breast cancer [59]. MiR-211 has been found to be reduced in most melanomas, and expression of mRNA and the protein KCNMA1 has an inverse correlation with miR-211. Melastatin is a tumor suppressor and is downregulated in metastatic melanomas. Its tumor suppressive activity is mediated through miR-211. It has been reported that miR-211 is hosted in melastatin intron [60].

2.2.4 Others (miR-1, miR-126, let-7, miR-29c and miR-214)

Recently, miR-1 expression has been shown to be lost in human primary lung cancer tissues and cell lines. In smooth muscle cells and cardiac muscles, miR-1 is highly expressed. In epithelial cell lines (A549 and H1299), miR-1 has reversed tumorigenic properties such as growth, replication potential, motility or migration, clonogenic survival, and tumor formation in nude mice. Expression of oncogenic molecules such as a receptor tyrosine kinase, the MET, Pim-1 (a Ser/Thr kinase), and FoxP1 (a transcription factor) is diminished by miR-1. Also, expression induces apoptosis in A549 lung cancer cells in response to doxorubicin treatment of MiR-1 with enhanced activation of caspase 3, caspase 7, and cleavage of PARP-1 [61]. MiR-1 inhibits migration either by blocking the function of some oncogenes or by activating genes in the apoptosis signaling pathway.

MiR-126 has been identified as a putative target of Crk in nonsmall cell lung carcinoma cell lines [62]. Crk, or v-crk sarcoma virus CT10 oncogene homolog, is a member of a family of adaptor proteins that are involved in intracellular signal pathways and play an important role in cell adhesion, proliferation, and migration. MiR-126 modifies lung cancer cell phenotype by inhibiting adhesion, migration, and invasion through Crk regulation. This demonstrates an important role of miR-126 in migration, invasion, and adhesion of nonsmall cell lung carcinoma cells [62].

Raf kinase inhibitory protein (RKIPor PEBP1) is a member of the evolutionarily conserved phosphatidylethanolamine binding protein family. It negatively regulates G- protein-coupled receptor kinase-2, as well as the MAP kinase (MAPK) and NF-κB signaling cascades. Metastatic progression is suppressed by RKIP without affecting primary tumor growth in an orthotopic murine model of androgen-independent prostate tumors. RKIP is critical in the induction of let-7/miR-98. It also represses invasion, intravasation, and bone metastasis of breast tumor cells through a signaling cascade involving inhibition of MAPK, Myc, and LIN28. This leads to the induction of microRNA let-7 and downregulation of its target genes [63, 64]. Thus, understanding the biogenesis of miRNAs will be useful in identifying novel therapeutic targets.

MiRNAs also have major roles in highly invasive tumor nasopharyngeal carcinoma (NPC). MiR-29c is down-regulated in nasopharyngeal carcinoma. Most of the miR-29c-targeted genes encode extracellular matrix proteins, including multiple collagens and laminin γ1. These proteins are associated with tumor cell invasion and metastasis [65]. MiR-29c is an important target to develop novel clinical strategies to suppress NPC.

MiR-214 is overexpressed in metastatic melanoma cell lines and tumor specimens, and regulates the expression of two transcription factors, AP-2γ and AP-2α. These factors have been shown to be important in melanoma invasion, metastasis, and angiogenesis and is thought to be a driver of melanoma metastasis [66].

2.3 *miRNAs Suppress Tumor Metastasis*

2.3.1 miR-335, miR-206 and miR-126

Through microRNA array profiling of metastatic breast cancer cells, it was determined that miR-126, miR-206, and miR-335 have important functions in cancer. Overexpression of miR-335, miR-206, or miR-126 inhibits metastatic activities. Specifically, expression of miR-335 and miR-126 is lost in most primary breast tumor patients. Loss of the miRNAs is associated with poor chances of metastasis-free survival. Tenascin C (TNC) and the SRY-box containing transcription factor SOX4 are the targets for miR-335. These studies suggest that miR-335, miR-126, and miR-206 have metastatic suppressor function in breast cancer [67]. Since the loss of microRNAs is strongly associated with metastatic relapse, these molecules may be used as prognostic markers for advanced breast cancer.

2.3.2 miR-31

Metastatic suppressors are natural regulators of cancer metastasis. miR-31 was discovered as an anti-metastatic miRNA [68]. Its attenuated expression was observed in metastatic breast cancer cells, and it acts at multiple steps of the invasion-metastasis cascade. The overexpression of miR-31 in MCF7-Ras cells has no effect on primary tumor growth and cell proliferation, but lung metastasis is enhanced. Some important functional genes, including myosin phosphatase-Rho interacting protein (MRIP), matrix metallopeptidase 16 (MMP16), radixin (RDX), frizzled3 (Fzd3), integrin α5 (ITGA5), and RhoA have been found to be repressed in miR-31 overexpressed MDA-MB231 breast cancer cells. It has been suggested that miR-31 is a key miRNA in the inhibition of metastasis through several mechanisms [68]. The miRNAs that have pleiotropic effects are particularly interesting, because all target genes might produce a cumulative effect, which is required to attack multifactorial diseases, including cancer. MiR-31 also functions as an oncogene by targeting tumor suppressors, including large tumor suppressor 2 and PP2A regulatory subunitB alpha isoform. MiR-31 and its target mRNAs have been found to be inversely expressed in mouse and human lung cancers [69].

2.3.3 miR-146a and miR-146b

MiR-146, which is important in regulating the nuclear factor-B pathway, is abundantly expressed in breast cancer metastasis suppressor1 (BRMS1)-expressing cells. BRMS1 is a nuclear protein that regulates multiple genes to suppress tumor metastasis without affecting orthotopic tumor growth in murine and human cancer cells. Transduction of miR-146a or miR-146b into MDA-MB-231 breast cancer cells downregulates expression of epidermal growth factor receptor, inhibits invasion and migration *in vitro*, and suppresses lung metastasis *in vivo* [70]. It has also been demonstrated that overexpression of miR-146b significantly reduces invasion and migration of glioblastoma U373 cells. MMP16 emerged as a critical target of miR-146b in regulating the migration of glioblastoma cells [71]. MiR-146a and miR-146b may be important in tumor suppression function and accordingly are good candidates for cancer therapeutics.

2.3.4 miR-205

Wu et al. (2009) found that miR-205 is highly expressed in normal breast tissues, nonmalignant breast epithelial cells, and MCF-10A cells. Ectopic expression of miR-205 in MCF7 significantly reduces cellular proliferation, clonogenic survival, and anchorage-independent growth. In MDA-MB-231 breast cancer cells, invasion and metastatic abilities are suppressed by miR-205 *in vivo*. Further, two direct targets of miR-205, ErbB3 and VEGF-A, have been found in breast cancer cell lines. Their expression is specifically suppressed by miR-205 through interaction with their binding site on 3′-UTR [72].

Compared to its undetectable expression levels in both androgen-dependent (VCaP, LNCaP) and androgen-independent (DU145, PC-3) prostate cancer cells, miR-205 is highly expressed in normal prostate tissues. Overexpression of miR-205 in prostate cancer cells elevates the expression of E-cadherin and reduces invasion, leading to mesenchymal epithelial transition (MET). Bioinformatics analysis has shown the miR-205 targets N-chimaerin, ErbB3, E2F1, E2F5, ZEB2, and protein kinase-C epsilon. However, these targets further need to be validated [73].

2.3.5 Others (miR-29b, miR-198, miR-34a, let-7f)

MicroRNA-29b can suppress tumor angiogenesis, invasion, and metastasis by regulating MMP-2 expression in hepatocellular carcinoma (HCC) [74]. MiR-198 has also been identified as a novel suppressor of HCC cell invasion by negative regulation of the HGF/c-MET pathway [75]. MiR-34a has been found to have multiple tumor suppressive effects in murine hepatocarcinoma, not only inhibiting cell growth by cell-cycle arrest, but also by repressing metastasis, and possibly serving as a novel therapeutic target for hepatocarcinoma [76]. Another potential metastasis inhibitor, microRNA let-7f, has been shown to inhibit tumor invasion and metastasis by targeting MYH9 in human gastric cancer [77].

3 Conclusion and Future Perspective

An explosion of interest in miRNAs is currently occurring. Many discoveries opened the floodgates of information to the understanding that miRNAs can function as either tumor suppressors or oncogenes, based on what kind of genes they target. The problem is that the vast majority of metastasis-related miRNAs have been identified in specific tumor types. Their functions need to be confirmed in many other robust metastatic cell types or animal models. Another problem is that some miRNAs have multiple functions. Before they can be used as therapeutic molecules or targets to treat various diseases, including cancer, they need to undergo rigorous systemic functional analysis. MiRNAs are also emerging as prognostic markers in the realms of cancer diagnosis. Since one miRNA can regulate several target proteins, and present studies of miRNA targets are based on computer softwares, the daunting challenge now facing biologists is to validate and confirm the roles of miRNAs in transgenic animal models. Studies with miRNAs may be useful for early prognosis. Moreover, miRNAs may serve as markers for therapy indicators. Studies with miRNAs may lead to the development of miRNA- based therapy either by miRNA suppression or by using miRNA like compounds. In the last decade, miRNA research has grown remarkably. Hence, we are very close to developing novel miRNA-targeted therapies and using miRNAs as diagnostic and prognostic markers.

Acknowledgments I would like to thank Samthosh Alahari for editorial help, and members of my lab for intellectual help. This work was supported by grants from NIH, Susan Komen Foundation and Louisiana Cancer Research Consortium.

References

1. Lee RC, Feinbaum RL, Ambros V (1993) The C. elegans heterochronic gene lin-4 encodes small RNAs with antisense complementarity to lin-14. Cell 75:843–854
2. Reinhart BJ, Slack FJ, Basson M, Pasquinelli AE, Bettinger JC, Rougvie AE, Horvitz HR, Ruvkun G (2000) The 21-nucleotide let-7 RNA regulates developmental timing in caenorhabditis elegans. Nature 403:901–906
3. Pasquinelli AE, Reinhart BJ, Slack F, Martindale MQ, Kuroda MI, Maller B, Hayward DC, Ball EE, Degnan B, Muller P, Spring J, Srinivasan A, Fishman M, Finnerty J, Corbo J, Levine M, Leahy P, Davidson E, Ruvkun G (2000) Conservation of the sequence and temporal expression of let-7 heterochronic regulatory RNA. Nature 408:86–89
4. Lagos-Quintana M, Rauhut R, Lendeckel W, Tuschl T (2001) Identification of novel genes coding for small expressed RNAs. Science 294:853–858
5. Lee RC, Ambros V (2001) An extensive class of small RNAs in caenorhabditis elegans. Science 294:862–864
6. Lau NC, Lim LP, Weinstein EG, Bartel DP (2001) An abundant class of tiny RNAs with probable regulatory roles in caenorhabditis elegans. Science 294:858–862
7. Calin GA, Dumitru CD, Shimizu M, Bichi R, Zupo S, Noch E, Aldler H, Rattan S, Keating M, Rai K, Rassenti L, Kipps T, Negrini M, Bullrich F, Croce CM (2002) Frequent deletions and

down-regulation of micro- RNA genes miR15 and miR16 at 13q14 in chronic lymphocytic leukemia. Proc Natl Acad Sci U S A 99:15524–15529
8. Heneghan HM, Miller N, Lowery AJ, Sweeney KJ, Kerin MJ (2009) MicroRNAs as Novel Biomarkers for Breast Cancer. J Oncol 2009:950201
9. Kan JS, Delassus GS, D'Souza KG, Hoang S, Aurora R, Eliceiri GL (2010) Modulators of cancer cell invasiveness. J Cell Biochem 111:791–796
10. Shi M, Liu D, Duan H, Shen B, Guo N (2010) Metastasis-related miRNAs, active players in breast cancer invasion, and metastasis. Cancer Metastasis Rev 29:785–799
11. Di Leva G, Croce CM (2010) Roles of small RNAs in tumor formation. Trends Mol Med 16:257–267
12. Bagga S, Bracht J, Hunter S, Massirer K, Holtz J, Eachus R, Pasquinelli AE (2005) Regulation by let-7 and lin-4 miRNAs results in target mRNA degradation. Cell 122:553–563
13. Inui M, Martello G, Piccolo S (2010) MicroRNA control of signal transduction. Nat Rev Mol Cell Biol 11:252–263
14. Yu Z, Willmarth NE, Zhou J, Katiyar S, Wang M, Liu Y, McCue PA, Quong AA, Lisanti MP, Pestell RG (2010) microRNA 17/20 inhibits cellular invasion and tumor metastasis in breast cancer by heterotypic signaling. Proc Natl Acad Sci U S A 107:8231–8236
15. Uziel T, Karginov FV, Xie S, Parker JS, Wang YD, Gajjar A, He L, Ellison D, Gilbertson RJ, Hannon G, Roussel MF (2009) The miR-17~92 cluster collaborates with the Sonic Hedgehog pathway in medulloblastoma. Proc Natl Acad Sci U S A 106:2812–2817
16. He L, He X, Lowe SW, Hannon GJ (2007) microRNAs join the p53 network–another piece in the tumour-suppression puzzle. Nat Rev Cancer 7:819–822
17. Garofalo M, Di Leva G, Romano G, Nuovo G, Suh SS, Ngankeu A, Taccioli C, Pichiorri F, Alder H, Secchiero P, Gasparini P, Gonelli A, Costinean S, Acunzo M, Condorelli G, Croce CM (2009) miR-221 & 222 regulate TRAIL resistance and enhance tumorigenicity through PTEN and TIMP3 downregulation. Cancer Cell 16:498–509
18. Pillai RS (2005) MicroRNA function: multiple mechanisms for a tiny RNA? Rna 11:1753–1761
19. Koturbash I, Zemp FJ, Pogribny I, Kovalchuk O (2010) Small molecules with big effects: the role of the microRNAome in cancer and carcinogenesis. Mutat Res 722:94–105
20. White NM, Fatoohi E, Metias M, Jung K, Stephan C, Yousef GM (2010) Metastamirs: a stepping stone towards improved cancer management. Nat Rev Clin Oncol 8:75–84
21. Dykxhoorn DM (2010) MicroRNAs and metastasis: little RNAs go a long way. Cancer Res 70:6401–6406
22. Baranwal S, Alahari SK (2009) miRNA control of tumor cell invasion and metastasis. Int J Cancer 126:1283–1290
23. Wu WK, Lee CW, Cho CH, Fan D, Wu K, Yu J, Sung JJ (2010) MicroRNA dysregulation in gastric cancer: a new player enters the game. Oncogene 29:5761–5771
24. Uhlmann S, Zhang JD, Schwager A, Mannsperger H, Riazalhosseini Y, Burmester S, Ward A, Korf U, Wiemann S, Sahin O (2010) miR-200bc/429 cluster targets PLCgamma1 and differentially regulates proliferation and EGF-driven invasion than miR-200a/141 in breast cancer. Oncogene 29:4297–4306
25. Li XF, Yan PJ, Shao ZM (2009) Downregulation of miR-193b contributes to enhance urokinase-type plasminogen activator (uPA) expression and tumor progression and invasion in human breast cancer. Oncogene 28:3937–3948
26. Reis PP, Tomenson M, Cervigne NK, Machado J, Jurisica I, Pintilie M, Sukhai MA, Perez-Ordonez B, Grenman R, Gilbert RW, Gullane PJ, Irish JC, Kamel-Reid S (2010) Programmed cell death 4 loss increases tumor cell invasion and is regulated by miR-21 in oral squamous cell carcinoma. Mol Cancer 9:238
27. Li Y, Zhang M, Chen H, Dong Z, Ganapathy V, Thangaraju M, Huang S (2010) Ratio of miR-196s to HOXC8 messenger RNA correlates with breast cancer cell migration and metastasis. Cancer Res 70:7894–7904

28. Su X, Chakravarti D, Cho MS, Liu L, Gi YJ, Lin YL, Leung ML, El-Naggar A, Creighton CJ, Suraokar MB, Wistuba I, Flores ER (2010) TAp63 suppresses metastasis through coordinate regulation of Dicer and miRNAs. Nature 467:986–990
29. Patel JB, Appaiah HN, Burnett RM, Bhat-Nakshatri P, Wang G, Mehta R, Badve S, Thomson MJ, Hammond S, Steeg P, Liu Y, Nakshatri H (2010) Control of EVI-1 oncogene expression in metastatic breast cancer cells through microRNA miR-22. Oncogene 30:1290–1301
30. Xu D, Takeshita F, Hino Y, Fukunaga S, Kudo Y, Tamaki A, Matsunaga J, Takahashi RU, Takata T, Shimamoto A, Ochiya T, Tahara H (2011) miR-22 represses cancer progression by inducing cellular senescence. J Cell Biol 193:409–424
31. Meng Z, Fu X, Chen X, Zeng S, Tian Y, Jove R, Xu R, Huang W (2010) miR-194 is a marker of hepatic epithelial cells and suppresses metastasis of liver cancer cells in mice. Hepatology 52:2148–2157
32. Su JL, Chen PB, Chen YH, Chen SC, Chang YW, Jan YH, Cheng X, Hsiao M, Hung MC (2010) Downregulation of microRNA miR-520h by E1A contributes to anticancer activity. Cancer Res 70:5096–5108
33. Baffa R, Fassan M, Volinia S, O'Hara B, Liu CG, Palazzo JP, Gardiman M, Rugge M, Gomella LG, Croce CM, Rosenberg A (2009) MicroRNA expression profiling of human metastatic cancers identifies cancer gene targets. J Pathol 219:214–221
34. Elson-Schwab I, Lorentzen A, Marshall CJ (2010) MicroRNA-200 family members differentially regulate morphological plasticity and mode of melanoma cell invasion. PLoS One 5(10):e13176
35. Sachdeva M, Mo YY (2010) MicroRNA-145 suppresses cell invasion and metastasis by directly targeting mucin 1. Cancer Res 70:378–387
36. Yu Z, Baserga R, Chen L, Wang C, Lisanti MP, Pestell RG (2010) microRNA, cell cycle, and human breast cancer. Am J Pathol 176:1058–1064
37. Iorio MV, Ferracin M, Liu CG, Veronese A, Spizzo R, Sabbioni S, Magri E, Pedriali M, Fabbri M, Campiglio M, Menard S, Palazzo JP, Rosenberg A, Musiani P, Volinia S, Nenci I, Calin GA, Querzoli P, Negrini M, Croce CM (2005) MicroRNA gene expression deregulation in human breast cancer. Cancer Res 65:7065–7070
38. Korpal M, Kang Y (2008) The emerging role of miR-200 family of microRNAs in epithelial-mesenchymal transition and cancer metastasis. RNA Biol 5:115–119
39. Ma L, Reinhardt F, Pan E, Soutschek J, Bhat B, Marcusson EG, Teruya-Feldstein J, Bell GW, Weinberg RA (2010) Therapeutic silencing of miR-10b inhibits metastasis in a mouse mammary tumor model. Nat Biotechnol 28:341–347
40. Anand S, Majeti BK, Acevedo LM, Murphy EA, Mukthavaram R, Scheppke L, Huang M, Shields DJ, Lindquist JN, Lapinski PE, King PD, Weis SM, Cheresh DA (2010) MicroRNA-132-mediated loss of p120RasGAP activates the endothelium to facilitate pathological angiogenesis. Nat Med 16 909–914
41. Gebeshuber CA, Zatloukal K, Martinez J (2009) miR-29a suppresses tristetraprolin, which is a regulator of epithelial polarity and metastasis. EMBO Rep 10:400–405
42. Gregory PA, Bert AG, Paterson EL, Barry SC, Tsykin A, Farshid G, Vadas MA, Khew-Goodall Y, Goodall GJ (2008) The miR-200 family and miR-205 regulate epithelial to mesenchymal transition by targeting ZEB1 and SIP1. Nat Cell Biol 10:593–601
43. Park SM, Gaur AB, Lengyel E, Peter ME (2008) The miR-200 family determines the epithelial phenotype of cancer cells by targeting the E-cadherin repressors ZEB1 and ZEB2. Genes Dev 22:894–907
44. Bracken CP, Gregory PA, Khew-Goodall Y, Goodall GJ (2009) The role of microRNAs in metastasis and epithelial-mesenchymal transition. Cell Mol Life Sci 66:1682–1699
45. Vetter G, Saumet A, Moes M, Vallar L, Le Bechec A, Laurini C, Sabbah M, Arar K, Theillet C, Lecellier CH, Friederich E (2010) miR-661 expression in SNAI1-induced epithelial to mesenchymal transition contributes to breast cancer cell invasion by targeting Nectin-1 and StarD10 messengers. Oncogene 29:4436–4448

46. Cao Q, Yu J, Dhanasekaran SM, Kim JH, Mani RS, Tomlins SA, Mehra R, Laxman B, Cao X, Yu J, Kleer CG, Varambally S, Chinnaiyan AM (2008) Repression of E-cadherin by the polycomb group protein EZH2 in cancer. Oncogene 27:7274–7284
47. Varambally S, Cao Q, Mani RS, Shankar S, Wang X, Ateeq B, Laxman B, Cao X, Jing X, Ramnarayanan K, Brenner JC, Yu J, Kim JH, Han B, Tan P, Kumar-Sinha C, Lonigro RJ, Palanisamy N, Maher CA, Chinnaiyan AM (2008) Genomic loss of microRNA-101 leads to overexpression of histone methyltransferase EZH2 in cancer. Science 322:1695–1699
48. Krichevsky AM, Gabriely G (2009) miR-21: a small multi-faceted RNA. J Cell Mol Med 13:39–53
49. Du J, Yang S, An D, Hu F, Yuan W, Zhai C, Zhu T (2009) BMP-6 inhibits microRNA-21 expression in breast cancer through repressing deltaEF1 and AP-1. Cell Res 19:487–496
50. Huang et al (2009) Up-regulating miR-21 by Her2/Neu signaling. J. Biol Chem 284:18515–18524
51. Li T, Li D, Sha J, Sun P, Huang Y (2009) MicroRNA-21 directly targets MARCKS and promotes apoptosis resistance and invasion in prostate cancer cells. Biochem Biophys Res Commun 383:280–285
52. Meng F, Henson R, Wehbe-Janek H, Ghoshal K, Jacob ST, Patel T (2007) MicroRNA-21 regulates expression of the PTEN tumor suppressor gene in human hepatocellular cancer. Gastroenterology 133:647–658
53. Hatley ME, Patrick DM, Garcia MR, Richardson JA, Bassel-Duby R, van Rooij E, Olson EN (2010) Modulation of K-Ras-dependent lung tumorigenesis by MicroRNA-21. Cancer Cell 18:282–293
54. Gabriely G, Wurdinger T, Kesari S, Esau CC, Burchard J, Linsley PS, Krichevsky AM (2008) MicroRNA 21 promotes glioma invasion by targeting matrix metalloproteinase regulators. Mol Cell Biol 28:5369–5380
55. Ma L, Teruya-Feldstein J, Weinberg RA (2007) Tumour invasion and metastasis initiated by microRNA-10b in breast cancer. Nature 449:682–688
56. Segura MF, Hanniford D, Menendez S, Reavie L, Zou X, Alvarez-Diaz S, Zakrzewski J, Blochin E, Rose A, Bogunovic D, Polsky D, Wei J, Lee P, Belitskaya-Levy I, Bhardwaj N, Osman I, Hernando E (2009) Aberrant miR-182 expression promotes melanoma metastasis by repressing FOXO3 and microphthalmia-associated transcription factor. Proc Natl Acad Sci U S A 106:1814–1819
57. Huynh C, Segura MF, Gaziel-Sovran A, Menendez S, Darvishian F, Chiriboga L, Levin B, Meruelo D, Osman I, Zavadil J, Marcusson EG, Hernando E (2010) Efficient *in vivo* microRNA targeting of liver metastasis. Oncogene 30:1481–1488
58. Wang G, Mao W, Zheng S (2008) MicroRNA-183 regulates Ezrin expression in lung cancer cells. FEBS Lett 582:3663–3668
59. Lowery AJ, Miller N, Dwyer RM, Kerin MJ (2010) Dysregulated miR-183 inhibits migration in breast cancer cells. BMC Cancer 10:502
60. Levy C, Khaled M, Iliopoulos D, Janas MM, Schubert S, Pinner S, Chen PH, Li S, Fletcher AL, Yokoyama S, Scott KL, Garraway LA, Song JS, Granter SR, Turley SJ, Fisher DE, Novina CD (2010) Intronic miR-211 assumes the tumor suppressive function of its host gene in melanoma. Mol Cell 40:841–849
61. Nasser MW, Datta J, Nuovo G, Kutay H, Motiwala T, Majumder S, Wang B, Suster S, Jacob ST, Ghoshal K (2008) Down-regulation of micro-RNA-1 (miR-1) in lung cancer. Suppression of tumorigenic property of lung cancer cells and their sensitization to doxorubicin-induced apoptosis by miR-1. J Biol Chem 283:33394–33405
62. Crawford M, Brawner E, Batte K, Yu L, Hunter MG, Otterson GA, Nuovo G, Marsh CB, Nana-Sinkam SP (2008) MicroRNA-126 inhibits invasion in non-small cell lung carcinoma cell lines. Biochem Biophys Res Commun 373:607–612
63. Fu Z, Smith PC, Zhang L, Rubin MA, Dunn RL, Yao Z, Keller ET (2003) Effects of raf kinase inhibitor protein expression on suppression of prostate cancer metastasis. J Natl Cancer Inst 95:878–889

64. Dangi-Garimella S, Yun J, Eves EM, Newman M, Erkeland SJ, Hammond SM, Minn AJ, Rosner MR (2009) Raf kinase inhibitory protein suppresses a metastasis signalling cascade involving LIN28 and let-7. Embo J 28:347–358
65. Sengupta S, den Boon JA, Chen IH, Newton MA, Stanhope SA, Cheng YJ, Chen CJ, Hildesheim A, Sugden B, Ahlquist P (2008) MicroRNA 29c is down-regulated in nasopharyngeal carcinomas, up-regulating mRNAs encoding extracellular matrix proteins. Proc Natl Acad Sci U S A 105:5874–5878
66. Penna E, Orso F, Cimino D, Tenaglia E, Lembo A, Quaglino E, Poliseno L, Haimovic A, Osella-Abate S, De Pitta C, Pinatel E, Stadler MB, Provero P, Bernengo MG, Osman I (2011) Taverna D microRNA-214 contributes to melanoma tumour progression through suppression of TFAP2C. Embo J 30:1990–2007
67. Tavazoie SF, Alarcon C, Oskarsson T, Padua D, Wang Q, Bos PD, Gerald WL, Massague J (2008) Endogenous human microRNAs that suppress breast cancer metastasis. Nature 451:147–152
68. Valastyan S, Reinhardt F, Benaich N, Calogrias D, Szasz AM, Wang ZC, Brock JE, Richardson AL, Weinberg RA (2009) A pleiotropically acting microRNA, miR-31, inhibits breast cancer metastasis. Cell 137:1032–1046
69. Liu X, Sempere LF, Ouyang H, Memoli VA, Andrew AS, Luo Y, Demidenko E, Korc M, Shi W, Preis M, Dragnev KH, Li H, Direnzo J, Bak M, Freemantle SJ, Kauppinen S, Dmitrovsky E (2010) MicroRNA-31 functions as an oncogenic microRNA in mouse and human lung cancer cells by repressing specific tumor suppressors. J Clin Invest 120:1298–1309
70. Hurst DR, Edmonds MD, Scott GK, Benz CC, Vaidya KS, Welch DR (2009) Breast cancer metastasis suppressor 1 up-regulates miR-146, which suppresses breast cancer metastasis. Cancer Res 69:1279–1283
71. Xia H, Qi Y, Ng SS, Chen X, Li D, Chen S, Ge R, Jiang S, Li G, Chen Y, He ML, Kung HF, Lai L, Lin MC (2009) microRNA-146b inhibits glioma cell migration and invasion by targeting MMPs. Brain Res 1269:158–165
72. Wu H, Zhu S, Mo YY (2009) Suppression of cell growth and invasion by miR-205 in breast cancer. Cell Res 19:439–448
73. Gandellini P, Folini M, Longoni N, Pennati M, Binda M, Colecchia M, Salvioni R, Supino R, Moretti R, Limonta P, Valdagni R, Daidone MG, Zaffaroni N (2009) miR-205 Exerts tumor-suppressive functions in human prostate through down-regulation of protein kinase Cepsilon. Cancer Res 69:2287–2295
74. Fang JH, Zhou HC, Zeng C, Yang J, Liu Y, Huang X, Zhang JP, Guan XY, Zhuang SM (2011) MicroRNA-29b suppresses tumor angiogenesis, invasion and metastasis by regulating MMP-2 expression. Hepatology 54:1729–1740
75. Tan S, Li R, Ding K, Lobie PE, Zhu T (2011) miR-198 inhibits migration and invasion of hepatocellular carcinoma cells by targeting the HGF/c-MET pathway. FEBS Lett 585 2229–2234
76. Guo Y, Li S, Qu J, Wang S, Dang Y, Fan J, Yu S, Zhang J (2011) miR-34a inhibits lymphatic metastasis potential of mouse hepatoma cells. Mol Cell Biochem 354 275–282
77. Liang S, He L, Zhao X, Miao Y, Gu Y, Guo C, Xue Z, Dou W, Hu F, Wu K, Nie Y, Fan D (2011) MicroRNA let-7f inhibits tumor invasion and metastasis by targeting MYH9 in human gastric cancer. PLoS One 6 e18409

MicroRNA in Leukemias

Deepa Sampath

Abstract MicroRNAs are a class of master gene regulators that are extensively deregulated in all leukemias. Distinctive miRNA signatures are expressed in each cytogenetically defined subtype of leukemia which play multifaceted roles in the genesis, pathobiology, progression and clinical outcome of this group of diseases. This chapter aims to provide a comprehensive review of the expression, regulation and cellular targets of the microRNAs in acute lymphoblastic leukemia (ALL), acute myeloid leukemia (AML), chronic lymphocytic leukemia (CLL) and chronic myeloid leukemia (CML) in order to understand the regulatory role of microRNA in the leukemias as well as to identify novel therapeutic targets.

1 Introduction

Leukemia is a malignant disorder of the blood and bone marrow that is characterized by the clonal accumulation of myeloid or lymphoid cells that become arrested in varying stages of the hematopoietic maturation process. Myeloid and lymphoid leukemia can manifest in either an acute or chronic form; in acute lymphoblastic leukemia (ALL) and acute myelogenous leukemia (AML) dysfunctional lymphoblasts or myeloblasts accumulate by uncontrolled proliferation that can be fatal within weeks or months, if untreated. In chronic leukemias, the onset is slow and abnormal lymphocytes or myelocytes accumulate over many months or years to result in chronic lymphoblastic leukemia (CLL) and chronic myelogenous leukemia (CML) respectively. All leukemias harbor numerous chromosomal and genetic aberrations that are the single most determinants of clinical outcome [29, 40, 57, 58, 85, 91, 95]. More recently, it was discovered that a novel class of gene regulators, the microRNAs, were causally linked to cancer [16–18, 37]. This discovery led to the extensive profiling of miRNA expression in all cancers, including leukemia

D. Sampath (✉)
Department of Experimental Therapeutics, University of Texas M.D. Anderson Cancer Center, 1515 Holcombe Blvd, Box 71, Houston, TX 77030, USA
e-mail: dsampath@mdanderson.org

S. Alahari (ed.), *MicroRNA in Cancer*,
DOI 10.1007/978-94-007-4655-8_7, © Springer Science+Business Media Dordrecht 2013

[76]. Interestingly, it was found that each leukemia expressed unique signatures of microRNA that segregated along with specific cytogenetic or gene abnormalities with such fidelity that these miRNA signatures could be used to distinguish between subtypes of leukemias [76]. MicroRNAs are a class of non-coding RNA molecules 18–22 nucleotides long that function as master gene regulators in normal and cancer cells. MicroRNAs epigenetically alter gene expression by binding to complimentary sequences located on the 3′untranslated regions (3′UTR), or less frequently the 5′untranslated regions (5′UTR) and coding regions of target mRNAs to either mediate its degradation or inhibit its translation [3, 5, 6, 44, 120]. Consequently depending on cellular context and targets repressed microRNAs functioned either as tumor suppressors or oncogenes [37]. This chapter aims to provide a comprehensive review of the expression, regulation and cellular targets of the microRNAs in leukemia that may help identify new mechanisms that underlie the pathobiology of disease as well as new therapeutic targets.

1.1 MicroRNAs in ALL

1.1.1 MicroRNA Expression Profiling

ALL is the most common pediatric malignancy that accounts for about 75 % of all childhood leukemias but is rare in adults. ALL occurs due to the clonal proliferation of immature B or T lymphocytes and is an heterogeneous disease that is sub-classified based chromosomal, immunophenotypic and structural features [69]. Cytogenetic and molecular abnormalities occur in 90 % of ALL and are the most important predictors of clinical outcome. The first study to characterize microRNA expression in ALL conducted miRNA array analyses in pooled RNA samples from seven different subtypes of ALL and compared them to normal lymphocytes that had been positively selected for the CD19 cell surface marker. This study identified that miR-128b, miR-204, miR-218, miR-331, and miR-181b-1 and the miR-17-92 cluster were upregulated in ALL samples [128]. *However, since miRNA expression patterns recapitulate the cell lineages they arise from* [76], *and each subtype of ALL arises from distinct cell lineages, pooling of RNA from subtypes would identify alterations in miRNA expression depending on the type of normal cell used as a comparator. The use of CD19 purified lymphocytes as the normal comparator was ideal for comparing microRNA signatures of B-ALL as CD19 is expressed largely on B-lymphocytes, but it was not an appropriate comparator to evaluate miRNA profiles obtained from T-ALL as T-lymphocytes largely express CD3, not CD19. Secondly, pooling of RNA samples would again identify miRNA that were deregulated in any subtype of ALL but hamper the identification of miRNA expression signatures that were unique to individual subtypes of ALL.* A second study performed a similar analysis in CD19+ and pre-B-ALL and showed that miR-222, miR-339, and miR-142-3p were over-expressed in pre-B-ALL whereas hsa-miR-451 and hsa-miR-373* were down-regulated [62]. A third study compared ALL samples with CD34+ cells as a normal comparator. Using these criteria, 14 microRNA genes, miR-128a, miR-142-3p, miR-142-5p, miR-150, miR-181a, miR-181b, miR-181c, miR-193a, miR-196b, miR-30e-5p, miR-34b, miR-365, miR-582, miR-708 were up-regulated whereas five (miR-100, miR-125b, miR-151-5p, miR-99a, let-7e) were downregulated [115]. The next study compared microRNA expression in different subtypes of ALL, namely B vs T-ALL and B-ALL subtypes expressing BCR/ABL, E2A/PBX1 or MLL/AF4 and found that each of these expressed a distinct miRNA signature. *Most importantly, just 3 miRNA, miR-148, miR-151, and miR-424, were sufficient to accurately discriminate between B- and T-ALL whereas a set of six miRNAs, miR-425-5p, miR-191, miR-146b, miR-128, miR-629, and miR-126 was able to distinguish between specific molecular lesions within B-ALL subtypes* [49]. Together, all these studies highlighted that the major subtypes of ALL express unique microRNA signatures that differ from each other and from those in healthy hematopoietic cells. Recent technological advances such as sequence by synthesis and high through put sequencing have allowed unbiased and detailed identification of miRNA expressed in ALL. Collectively these studies identified over 1,000 miRNA genes in the different subtypes of ALL of which about 150 miRNA were known, 16 were novel and 170 were candidate small RNA that had features of miRNAs which were uniquely expressed in ALL in comparison to normal hematopoietic cells [117, 132]. Identification and cataloging of these gene signatures is the first step and will lay a foundation for future studies that will elucidate the functional roles and cellular targets of miRNAs in each subtype of ALL.

1.1.2 MiRNA Regulation, Cellular Function and Prognostic Significance

While gene profiling provides a direct measure of the expression levels of miRNA in cells and readily identifies genes that are expressed abnormally, complimentary approaches that assess DNA and chromatin modifications around gene promoters offer another powerful strategy by which to identify genes that are aberrantly regulated in cancer. DNA methylation is the process by which a methyl group is added to carbon 5 of the cytosine within the dinucleotide CpG around gene promoters. DNA methylation is usually associated with gene silencing. In parallel, the histones around gene promoters control gene expression by virtue of multiple posttranslational modifications such as acetylation, methylation, phosphorylation, ubiquitylation, and sumoylation that occur at their tails. These modifications work together to either facilitate or prevent access to transcriptional factors that activate transcription and dictate whether the histones around gene promoters have or lack trimethylation of lysine 4 on histone H3 (H3K4me3), a key modification that is closely linked to transcriptional activation. Lack of H3K4me3 is usually associated with gene silencing whereas the converse is associated with gene activation [34].

Hypermethylation of gene promoters is already known to be a common occurrence in ALL and is associated with the prognosis of the disease and response to therapy [30, 38, 59, 63, 74, 83, 109, 126]. By evaluating the lack of H3K4me3 around gene promoter CpG islands Roman-Gomez et al. identified that 13 miRNA were hyper-methylated in ALL (miR-124a1, miR-124a2, miR-124a3, miR-34b/c, miR-9-1, miR-9-3, miR-10b, miR-203, miR-196b, miR-9-2, miR-132/212). More significantly, patients who had methylation of one or more 1 miRNA had an inferior disease-free survival and overall survival of 24 and 28 % in comparison to patients who did not have methylation (78 and 71 % respectively) at 14 years. *While this suggests a strong role for methylation silencing of miRNA in the prognosis of ALL, it remains to be determined whether the patients whose leukemia cells displayed methylation silencing of miRNA also underwent methylation silencing of other genes that have already been associated with disease prognosis and survival in ALL.* Two parallel studies conducted around the same time focused on MLL-rearranged ALL. The first identified a unique signature that was characterized by extensive promoter hypermethylation of miRNA genes. Of the 11 miRNA that were down-regulated in association with promoter hypermethylation, 5 (including miR-152) targeted either MLL or the DNA methytransferase 1 gene [122]. The second study found 8 miRNA aberrantly expressed in MLL vs non-MLL ALL, of which miR-708 was expressed at 100–1000 fold higher and miR-196 was expressed at 500–800 fold higher levels in MLL rearranged-ALL [115]. Other miRNA that became epigenetically silenced were miR-203 [14, 32], miR124 [1], miR-22 [72] and miR34a [33]. Loss of expression of each of these miRNA contributed to the pathobiology of ALL in that miR-203 negatively regulated BCR/ABL [14, 32], miR-124 negatively regulated Cdk6 and its loss was associated with a poor prognosis [1], and miR-34a was a p53 regulated miRNA that facilitated the p53 apoptotic program [33, 107]. Together, these findings underscore the importance of epigenetic mechanisms regulated by DNA methylation or histone modifications in mediating the selective silencing of key miRNA so as to contribute to the pathobiology of ALL.

Other miRNA of importance in ALL are the miR-17~92 cluster, the miR-125b and miR-196. The miR-17~92 cluster encodes the miR-17, miR-18a, miR-19a, miR-19b, miR-20 and miR-92. These miRNA regulate E2F1 to suppress apoptosis in ALL [96]. Further analysis of the role of individual miRNA within the mir-17-92 cluster identified that miR-19 was sufficient to promote leukemogenesis in ALL. miR-19 targeted the PTEN, a negative regulator of the phospho-inositol-3 kinase (PI3K) pro-survival pathway and Bim a pro-apoptotic Bcl-2 family member to effect its leukemogenic action [84]. miR-125b is another miRNA that is expressed 90–100 fold higher levels in ALL bearing the t(11;14)(q24;q32) translocation and promotes the survival of ALL both by targeting p53 and in a p53-independent manner [12, 28, 55, 115]. Finally, miR-196b is known to be expressed at high levels in ALL. One study demonstrated that miR-196b functioned as a tumor suppressor and targeted c-myc. However since c-myc developed mutations in its 3′ untranslated region (3′UTR) it lost the ability to bind to miR-196b, became over-expressed in ALL and was able to promote tumorigenesis [9]. Finally, a combined expression profile based on 14 microRNA was individually associated with prognosis and highly predictive of clinical outcome in pediatric T-ALL [116]. The expression and physiological significance of miRNA important in ALL are presented in Table 1.

Table 1 MicroRNA of functional importance in ALL

MicroRNA	Significance	Reference
miR-128a, miR-128b, let7b and miR-223	Accurately identified ALL from AML	[87]
miR-148, miR-151, and miR-424	Distinguished B-ALL from T-ALL	[49]
-425-5p, miR-191, miR-146b, miR-128, miR-629, and miR-126	Dsitinguishes between molecular subtypes of ALL	[49]
miR-17~92 cluster	Overexpressed in ALL, targets E2F1	[96]
miR-19b, miR-20a, miR-26a, miR-92 and miR-223	Leukemogenic- highly expressed in T-ALL-Target many tumor suppressors in ALL	[84]
544 known, 28 novel and 431 candiate miRNA	First report of RNA-sequencing in ALL	[117]
miR-203	Tumor suppressor miRNA-hypermethylated in ALL-results in BCR-ABL over-expression	[14]
miR-124 family, miR34a, miR-9, miR-10b, miR-203, miR-196	Hypermethylated in ALL-negatively affects OS	[110]

2 MicroRNA in AML

AML, like ALL is a highly heterogeneous disease that is characterized by the excessive proliferation and/or differentiation arrest of immature myeloblasts. About 30 % of all AML have structural or chromosomal abnormalities which number over 200 by now and strongly influence clinical outcome [40, 80]. Cytogenetic abnormalities that involved chromosomal translocations such as t(8;21)(q22;q22) and inv16(p13q22)/t(16;16)(p13;q22) led to alterations in the core binding factor (CBF) and were generally associated with a favorable prognosis whereas monosomies of chromosome 5 and 7, deletions in chromosome 5 [del(5q)], abnormal chromosome 3 (3q), or complex cytogenetics were linked to a unfavorable prognosis. Other chromosomal abnormalities such as gain of chromosomes 8, 21, or 22, deletions in chromosome 7 [del(7q)], or 9 [del(9q)], abnormal 11q23, all other structural or numerical changes were considered to impart an intermediate prognosis [92–94]. An additional 40–50 % of AML were found to have a cytogenetically normal (CN-AML) profile but varied widely in their outcome [79]. The advent of molecular profiling identified a number of molecular aberrations that occurred in CN-AML and provided a molecular basis for the widely differing clinical outcomes in this type of AML [79]. For instance, molecular aberrations such as the FLT3 internal tandem duplication (FLT3-ITD) and FLT3 tyrosine kinase domain (FLT3-TKD) mutations, and others such as KIT, WT1 and MLL mutations predicted an unfavorable prognosis whereas mutations in CEBPA or NPM1 were associated with a favorable prognosis [81]. *The association between of miRNA expression and clinical outcome was revealed when studies showed that a small set of miRNA, miR-199a, miR-199b, miR-191, miR-25, miR-20a predicted a poor survival when over-expressed* [51] *whereas miR-29b and miR-181 predicted inferior survival when under-expressed*

[10, 52, 118]. The first study profiled miRNA in AML and identified 27 miRNAs that were differentially expressed between ALL and AML. In ALL, the expression of miR-128a and miR-128b was highly elevated whereas let-7b and miR-223 were down-regulated when compared to AML [87]. The next study evaluated miRNA signatures in 182 AML samples of varying subtypes in comparison to CD34+ normal lymphocytes and found that overall 26 miRNA were down-regulated in AML across subtypes [51]. *However, the most important finding of this and other miRNA expression profiling studies was that they unequivocally established that unique miRNA signatures were associated with each cytogenetic and molecular aberration-defined subtype of AML and could be used as accurate and sensitive diagnostic tools* [28, 39, 61, 66, 73].

2.1 AML with Cytogenetic Abnormalities

Thirty percent of AML undergo cytogenetic abnormalities, the most prevalent of which, such as t(8;21) and inv16(p13q22), lead to alterations in the in AML1 transcription factor which are collectively classified as core binding factor (CBF) AML. t(8:21) results in the juxtaposition of the DNA binding region of the AML1 (core binding factor) with the nuclear protein ETO to form the AML-ETO fusion protein. Since the ETO gene interacts with the histone deacetylase (HDAC) group of gene repressors, expression of the AML-ETO fusion protein results in the recruitment of HDAC and its co-repressors to genes regulated by AML1 leading to their aberrant silencing. One of the key genes to become silenced by AML-ETO was that of miR-223. Loss of miR-223 was associated with blocks in differentiation and leukemogenesis. Conversely ectopic expression of miR-223 or epigenetic re-expression of the gene was linked to cellular differentiation [45, 46]. Thus loss of miR-223 may offer a mechanism for the initiation of leukemia in CBF AML. A parallel study showed that in a subset of CBF AML, the AML-ETO silences the miR-221/222 cluster. *It remains to be determined whether loss of miR-223 and miR221/222 overlap in the same subsets of AML.* Loss of expression of this miRNA cluster leads to the over-expression of its cellular target, the KIT oncogene [13]. The presence of Kit mutations decreases the favorable prognosis normally associated with AML-ETO and appears to be prognostically important for survival in CBF AML [108]. In addition, CBF AMLs also demonstrate a 1–1000-fold lower levels of expression of miR-9, miR-196a, miR-10a, miR-135a, miR-125b, miR-148a and miR-133a. In view of the fact that AML-ETO functions as a gene repressor, silencing of miRNA genes is an expected phenomenon. However, it becomes counterintuitive when explaining the over-expression of miRNAs in this subtype of AML. For instance, miR-449b, miR-320, miR-126, miR-126* and miR-328 are expressed at high levels in t(8;21) AML [24, 71]. In particular, miR-126 targets the Polo like kinase (PLK1), a tumor suppressor, to promote the survival of AML cells and can be used to differentiate CBF AML from all other subtypes [73].

Inv16(p12q22) encodes an AML1 co-repressor that directly interferes with the ability of the AML1 core binding factor to transcriptionally regulate genes and converts it into a transcriptional repressor. Two miRNA, miR-149 and miR-29c were down-regulated in this subtype of AML [24]. Similarly, t(15;17) AML was accurately characterized by miR-224, miR-368, and miR-382 whereas MLL-rearranged AML was defines by high levels of expression of miR-126, miR126*, miR-194, miR-219, miR-224, miR-301,miR-326, miR-368, miR382 and miR-17-5p and miR-20a and low levels of miR-34b, miR-15a, miR-29a, miR-29c, miR-372, miR-30a, miR-29b, miR-30e, miR-196a, let-7f, miR-102, miR-331, miR-299, miR-193 [26, 50, 73, 78].

2.2 *AML with molecular abnormalities: CEBPA*

The transcription factor CAAT enhancer binding protein alpha (CEBPA) is deregulated by multiple mechanisms in AML [78, 79, 98, 99]. For instance, the AML-ETO fusion protein down-regulates CEBPA [98]; in parallel, CEBPA undergoes mutations in 10–20 % of AML [99]. CEBPA mutations often occur in cytogenetically normal AML (CN-AML) or in those with genomic losses of chromosome 7. The presence of CEPBA mutations was associated with a better prognosis for the CN-AML group. CEBPA functions as a transcription factor that promotes the activation of many microRNAs such as miR-223, miR-34a and the miR-29a-b cluster [42, 61, 78, 104]. Consequently, loss of CEBPA is associated with a global down-regulation of these miRNA. Of these, mir-34 A was found to target E2F3 and blocked myeloid cell proliferation [104]. Therefore, the CEPBA loss-mediated down-regulation of miR-34a may contribute to leukemogenesis [104]. *CEBPA mutations were also associated with the elevated expression of multiple genes and fifteen miRNA. Significantly, eight of the fifteen miRNA belonged to the miR-181 family. Presence of elevated levels of miR-181independently conferred a favorable prognosis in AML perhaps based on its ability to target the leukemogenic HOXA genes in AML* [118]. The other genes up-regulated were miR-128, miR-192, miR219-1-3p, miR-224, miR-335 and miR-340 [78].

2.3 *Nucleophosmin 1 Mutations*

Nucleophosmin mutations (NPM1) are insertions that occur in the C-terminal of nucleophosmin, that confer on it a nuclear localization signal and subsequent accumulation of NPM1 in the nucleus. NPM1 mutations are an early event in leukemogenesis and confer a favorable prognosis in AML especially when they occurred in the absence of alterations in FLT3. Conversely, wild type NPM1 is associated with an inferior outcome which is worsened when it occurs in conjunction with the FLT3-ITD. A striking characteristic of gene expression profiles in NPM1 mutated AML was the up-regulation of HOX genes and their embedded microRNAs such as miR-10a, miR-10b, miR-196a and miR-196b. Other miRNA that were over-expressed were miR-15-16 and miR-17-92 cluster [50] along with the let-7 family, miR-100 and miR-9. MiRNA that were down-regulated were miR-204 that target HOXA10 and MEIS1 [50], miR-126, miR-130a and miR-451 [7].

2.4 FLT3-ITD and Tyrosine Kinase Domain Mutations

FLT3 is a receptor tyrosine kinase that has an important role in the proliferation of hematopoetic stem cells. Mutations in FLT3 occur in 33 % of AML either due to an internal tandem duplication (ITD) in the FLT3 juxtamembrane domain or due to point mutations in the FLT3 tyrosine kinase domain (TKD). Presence of a FLT3-ITD is not only associated with inferior survival but also predicts a rapidly proliferating disease and higher risk of relapse. While some profiling studies did not find any specific miRNA signature associated with FLT3-ITD [61] others found that over-expression of miR-155 and miR-125b [50] and under-expression of miR-144 and miR-451, miR-488 and miR-486-5p were key features of FLT3-ITD AML [127]. *In a separate study, the same group found that the expression levels of a single microRNA, miR-181 had independent prognostic significance. Higher levels of miR-181 were linked to a favorable prognosis even in poor risk AML that displayed FLT3-ITD or wild type NPM1. Conversely, low levels of miR-181 were associated with poor prognosis.* [118].

2.5 MN1, BAALC and ERG

The transcriptional co-activator, MN1, is over expressed in several types of AML and particularly in CN-AML. Higher levels of MN1 is linked to a shorter overall survival and poor prognosis [25, 56]. MN1 over-expression was associated with a unique signature of fifteen miRNA of which 8 miRNA were over-expressed and included six members of the miR-126 family and miR-424. MN1 over-expression was also associated with the down-regulation of miR-16, miR-19a, miR-20a miR-100 and miR-196a [66, 67].

BAALC (Brain And Acute Leukemia, Cytoplasmic) is another gene whose expression predicts clinical outcome in CN-AML; higher levels are associated with inferior survival [4]. Elevated BAALC often occurs in conjunction with MN1 over-expression [66, 67]. High levels of BLLAC were associated with the up regulation of eight miRNA that included the six members of the miR-126 family, miR-130 and miR-222. High levels of BAALC were also associated with a down-regulation of miRNA within the HOXA cluster, namely, miR-10a, miR-10b, and miR-9 and let-7b [4].

The ETS-related gene, ERG, was initially discovered as a fusion partner with the FUS gene in AML that had t16;21)(p11;q22). ERG over-expression also occurred in AML of complex karyotypes and in a fraction of patients with CN-AML. High levels

Table 2 MicroRNA of functional importance in AML

MicroRNA	Associated cytogenetic or molecular aberration	Significance	Reference
Low miR-223, miR-221	AML-ETO	Block in differentiation, Overexpression of Kit oncogene	[13, 45, 46]
Low miR-149, 29c	Inv16(p12q22)	Overexpression of DNMT3a, DNMT3b, TCL1	[24]
High miR-181 family,	CEBPA mutations	Favorable prognosis, HOXA genes	[118]
High miR-10a-b, miR-196a-b	NPM1 mutations	Favorable prognosis	[50]
miR-181	FLT-3 ITD	Stratifies the poor prognosis of FLT3-ITD	[118]
High miR-126, miR-421 Low miR-16, miR-19a	MN1	Inferior survival, Targets in AML unknown	[66, 67]
Low miR-10a-b, miR-9, let-7	BAALC	Inferior survival	[4]

of ERG were also predictive of poor survival. ERG over-expression was associated with high levels of expression of miR-302d, a gene that maintains a stem cell phenotype in multiple systems. Conversely, ERG over expression was associated with low levels of miR-107, miR-148a, which is shown to target DNMT3B [119] and miR-196, which is predicted in silico to target ERG itself [36]. The expression and physiological significance of miRNA important in AML are summarized in Table 2.

3 MicroRNA in CLL

Chronic lymphocytic leukemia (CLL) is the most common form of leukemia in North America and Europe [64]. Although it has a small proliferative component (<1 %), the majority of neoplastic B-lymphocytes accumulate due to defects in apoptosis. CLL is also characterized by recurrent chromosomal aberrations that affect chromosomes 13 (del13q14), 12 (trisomy 12), 11(del11q) and 17 (del17p) [15]. Del11q and del17p lead to the loss of the ATM and p53 genes. In particular, loss of expression of p53 is associated with an abrogation of the mitochondrial pathway to apoptosis, lack of response to chemotherapy, aggressive disease and inferior survival [129, 130].

3.1 MicroRNA Expression Profiling

The first connection between miRNA expression and cancer was made by Calin and Croce when they found that two microRNA genes, miR-15a and miR-16 were located at the chr13q14 genomic locus that became deleted in over 55 % of chronic lymphocytic leukemia (CLL) [19]. This group also identified that over half of all miRNA genes were located in cancer-related genomic loci or fragile sites [21]. Calin and Croce then evaluated the expression of miRNA of a large cohort of CLL and normal CD5 purified B cells and *found that not only was CLL characterized by a global downregulation of miRNA* but that a unique signature differentiated CLL from normal B-cells [20]. This signature was characterized by low levels of expression of miR-183, miR-190, miR-24, miR-213, miR-223, and most importantly, miR15a and miR-16 [20]. *Interestingly, these signatures also revealed the complexity of miRNA regulation in disease. For instance, the miR-17-92 cluster is transcribed as a single unit that contains miR-17, miR-18, miR-19a, miR-19b, miR-20 and miR-92. Of these miR-19 and miR-92 are over-expressed in CLL cells whereas miR-17 and miR-20 are down-regulated* [20] *underscoring the importance of post transcriptional miRNA processing mechanisms in dictating the levels of mature miRNA in cells. Similarly, miR-15a and miR-16 are transcribed as a single unit but the levels of miR-15a and usually lower than miR-16 suggesting that the processing of miR-15a was defective* [19]. Calin and Croce then compared miRNA signatures in CLL samples that differed in their status of the zeta-chain (TCR)-associated protein kinase 70 kDa (ZAP70) [20]. The presence of ZAP70 was previously shown to predict inferior survival and progressive disease [105]. They found that certain miRNA that were expressed at normal levels in indolent or ZAP 70 negative became under-expressed in CLL that was ZAP 70 kinase positive and likely to have progressive disease Prominent among these miRNA were three members of the mir-29 family, miR-29a, miR-29b and miR-29c [22]. *Elucidation of the mechanism by which these miRNA become silenced in aggressive CLL would be very important to unraveling the puzzle of CLL progression.* While Calin and Croce used a custom miRNA oligonucleotide array to evaluate miRNA expression, other groups used cloning and quantitative RT-PCR to identify that miR-155, miR-21, miR-150 and miR-92 were over-expressed whereas miR-181, miR-30d, let-7a, miR-15a and miR-16 were under-expressed in CLL [48, 82, 128]. Other profiling studies evaluated the *in vivo* significance of the expression of specific miRNA and identified that low levels of expression of miR29c and miR-223 specifically segregated with CLL that expressed adverse prognosis markers irrespective of the marker (unmutated heavy chain, beta2 microglobulin expression, or high risk cytogenetics) chosen for comparison [121], *supporting the growing body of data that suggested that miRNA expression profiles had in vivo significance and improved disease stratification.* In support of this, others identified that miR34a, miR-29c and miR-17-5p were preferentially down-regulated in CLL that had deletions in 17p (loss of p53) [89, 90] and correlated with adverse prognosis. The next study identified the karyotype specific expression of miRNA where specific miRNA subsets accurately differentiated between the common chromosomal aberrations in CLL. For instance, cytogenetically normal CLL was characterized by robust expression of miR-148, del13q CLL over-expressed miR-640 but under-expressed miR-155, trisomy 12 CLL over-expressed miR-146 family members but under-expressed miR-148 and miR-640, del11q CLL under-expressed both miR-29b and miR-155 whereas del17p under-expressed miR-34a, miR29c and miR-151 [124]. With the advent of direct RNA sequencing, it is now evident that CLL expressed over 300 previously unknown miRNA in addition to the already characterized 386 miRNA [60]. A very recent report demonstrated that certain extracellular miRNAs are present in CLL patient plasma at levels significantly different from healthy controls and from patients affected by other hematologic malignancies demonstrating the feasibility of using plasma levels of miRNA to diagnose CLL [88]. The study also determined that the level of circulating miR-20a correlates reliably with diagnosis-to treatment time [88].

3.2 Regulation, Cellular Targets and Clinical Significance

The microRNA cluster, miR-15a and miR-16 functions as a tumor suppressor in CLL because loss of miR-15a and miR-16 or, even just miR-16 led to the spontaneous generation of CLL in mice [65, 106]. Conversely, ectopic expression of miR-15a and miR-16 targeted Bcl-2 [35] Mcl-1[23] and ZAP70 [43], induced apoptosis in cell lines [35] and suppressed tumorigenesis in xenograft leukemia [23] and solid tumor models [11]. miR-15a and miR-16 are located on an intron of the dleu2 gene at the 13q14 locus and are coordinately expressed along with their host gene (68). Deletions in chromosome 13 [del(13q14)] are the most common cytogenetic aberration in CLL, usually involve the loss of a single allele in over >50 % of patients and occasionally both alleles (24 %) [65], and result in genomic losses centered around the miR-15a and miR-16 locus [86]. However, low levels of expression of miR-15a and miR-16 are not always linked to del(13q14) because many samples that express low levels of these miRNA do so despite a lack of observable aberrations in chromosome 13[97]. Furthermore, in addition to the miR-16 transcribed in conjunction with dleu2, mature miR-16 is also derived from a related microRNA gene cluster, miR-15a-16-2, located on the Smc4 gene on chromosome 3q26[19]. However, smc4 is expressed at low levels in CLL [19]. *These findings suggest the existence of additional regulatory mechanisms that might function in conjunction with del(13q14) to mediate the silencing of miR-15a and miR-16 in CLL.* A very recent report identified that the tumor suppressor protein p53 functioned as a transcriptional activator of the dleu2-miR-15a-miR-16 gene; consequently del17p CLL that lack p53 function would be expected to express lower levels of miR-15a and miR-16 [43].

The miR-29 family of microRNAs is another miRNA cluster that has an important role in CLL. The miR-29 family consists of three members, miR29a, miR-29b and miR-29c that are located on chromosome 1 and 7. miR-29a and miR-29b were found expressed at elevated levels in comparison to B lymphocytes in CLL and some other cancers. Transgenic mice expressing miR-29a developed an indolent form of leukemia suggesting that this miRNA functioned as an oncogene [114]. Paradoxically, miR-29a, miR-29b and miR-29c became selectively down regulated in aggressive CLL [22] by unknown mechanisms. In fact, miR-29b is better described as a tumor suppressor in many tumor types based on its ability to target Mcl-1[53], SP1[75], DNMT3a and b [54], Tcl-1[106] and Cdk6 [133]. miR-29b co-operates with miR-181 to target Tcl-1. In addition to Tcl-1 miR-181 also targets Mcl-1, is down-regulated in aggressive CLL and predicts overall survival [125]. The expression of miR-181 is silenced by DNA hypermethylation in CLL, an event that promotes CLL survival [100].

Table 3 MicroRNA of functional importance in CLL

MicroRNA	Associated cytogenetic or molecular aberration	Significance	Reference
Low miR-15a-16-1	del13q14	Overexpression Bcl-2, Mcl-1, ZAP70, generation of CLL	[19, 65, 43]
Low miR-29a-b-c, miR-223	None	Overexpression of Mcl-1, Tcl-1, Progressive disease	[20, 121]
Low miR181	None	Overexpression of Tcl-1, Progressive disease	[125]
Low miR34a, miR-29c	del17p	Chemoresistance	[89]
Low miR-106b	None	Suppression of p73	[112]
High miR-155, miR-21	del17p	Stratified risk, poor overall survival	[111]
300 known, 386 new miRNA	RNA-seq	RNA sequencing identified new miRNA	[60]

The miR-34 family of miRNA is transactivated by p53 and functions within the p53 signaling network to induce apoptosis [27]. Consequently, it is expected that miR-34a is down-regulated in del17p CLL (p53 loss) [89, 130] However, miR-34 is also silenced by epigenetic mechanisms such as DNA hypermethylation [31]. The expression levels of miR-34a had *in vivo* significance, low levels predicted a shorter time to treatment and rapid lymphocyte doubling time [2], and predicted resistance to fludarabine and other DNA damaging agents in CLL [130, 131]. In addition to miR34a, there appeared to be additional miRNA that included miR-221/222 that could differentiate between fludarabine sensitive and resistant CLL samples [47] *suggesting that differential expression of these miRNA play a role in the sensitivity to this therapeutic agent in CLL.*

The miR-106-25-93 cluster illustrates the context dependent action of miRNA in target cells. miR-106b is over-expressed in solid tumors and lymphoma where it targets p21 and Bim to suppress apoptosis [102, 103]. However, in CLL the miR-106b cluster is silenced by the action of histone deacetylases [112]. Re-expression of miR-106b was linked to the targeting of the Itch, E3 ubiquitin ligase, and reciprocal accumulation of its ubiquitin target p73. *This report identified a mechanism by which the expression of p73 could be induced so as to induce apoptosis in place of p53 in CLL.* Other miRNA that were epigenetically silenced in CLL were miR-203, miR-181, miR-424 and miR-107 [100].

Finally, while most miRNA are under-expressed in CLL, two miRNA miR-155 and miR-21 are expressed at high levels in CLL. miR-155 is generally expressed in the proliferation centers as well as in activated CLL and may represent the dividing fraction in this largely indolent disease [70, 124]. miR-21 was preferentially over-expressed in del17p CLL appeared to identify patients with poor prognosis [111]. The expression and physiological significance of miRNA important in CLL are summarized in Table 3.

4 MicroRNA in CML

CML is a proliferative disorder of the myeloid cells in the bone marrow that is largely characterized by a reciprocal chromosomal translocation t(9;22) that juxtaposes the breakage cluster region (BCR) on chromosome 22 with the Abl gene on chromosome 9 to form a fusion protein called BCR-ABL. BCR-ABL is a constitutively activated tyrosine kinase that activates multiple downstream signaling networks such as those of the PI3 kinase-Akt survival pathway, the STAT3/5-Mcl-1-BclX$_L$ antiapoptotic pathway, the Cdk2 cell cycle protein and finally attenuation of DNA repair to promote the proliferation, clonal evolution and survival of CML cells. CML starts with an indolent phase that very few (5 % or less) chromosomal aberrations other than BCR-ABL, but advances to an accelerated phase with 30 % chromosomal aberrations that increase to 80 % in the blast phase of CML. The discovery of imatinib, a tyrosine kinase inhibitor that targeted BCR-ABL was the first example of a successful molecularly targeted therapy and heralded an era of targeted therapeutics.

However, despite its spectacular clinical success, resistance to imatinib occurs in 20–25 % of patients. Under these circumstances, it becomes imperative to identify the molecular predictors of resistance to imatinib therapy. The first to evaluate whether specific miRNA could predict resistance to imatinib compared miRNA expression in cells from CML patients who responded or were resistant to therapy. The study identified 19 miRNAs that were differentially expressed between resistant and responder samples: 18 of them were down-regulated (hsa-miR-7, hsa-miR-23a, hsa-miR-26a, hsa-miR-29a, hsa-miR-29c, hsa-miR-30b, hsa-miR-30c, hsa-miR-100, hsa-miR-126, hsa-miR-134, hsa-miR-141, hsa-miR-183, hsa-miR-196b, hsa-miR-199a, hsa-miR-224, hsa-miR-326, hsa-miR-422b and hsa-miR-520a) whereas one was up-regulated (hsa-miR-191) in resistant CML patients [113]. A second study identified that one miRNA in particular miR-181 was 10–20 fold down-regulated in imatinib resistant cells; loss of this miRNA led to high levels of expression of Mcl-1, a pro-survival protein [134]. A third study compared the miRNA expression profiles of 49 miRNAs in CML patients at diagnosis, in hematological relapse, therapy failure, blast crisis and major molecular response and found that miR-150, miR-20a, miR-17, miR-19a, miR-103, miR-144, miR-155, miR-181a, miR-221 and miR-222 were deregulated in CML. These miRNA targeted cell cycle proteins, mitogen activated kinase-like protein (MAPK), epidermal growth factor receptor (EGFR, ERBB), transforming growth factor beta (TGFB1), p53 and the MYB transcription factor [77]. BCR-ABL also mediated the over-expression of several miRNA such as, miR-130a, miR-130b, miR-148a, miR-212 and miR-425-5p that targeted CCN3 to promote CML cell proliferation [123].

While miRNA are classical known to bind to target mRNAs to inhibit their translation one study uncovered a novel mechanism of miRNA action in CML-blast crisis (CML-BC). During progression from indolent to blast phase, BCR-ABL induced the expression of hnRNP E2, which functioned to sequester CEBPA mRNA to prevent the translation of this master regulator of myeloid differentiation. This

Table 4 MicroRNA of functional importance in CML

MicroRNA	Significance	Reference
19 miRNA signature	Differentiates between responders/non responders to imatinib	[113]
Low miR-181	Overexpression of Mcl-1, resistance to imatinib	[134]
High miR-150, miR-20a, miR-17, miR-19a, miR-103, miR-144, miR-155, miR-181a, miR-221 and miR-222 miR34a, miR-29c	Targets multiple tumor suppressor proteins	[77]
miR-328	Targets Pim kinase and Functions as a molecular decoy that binds to a repressor of CEBPA, releasing CEBPA to promote differentiation of myeloid cells	[41]
Low miR-328	Silenced by HDACs, promotes CML	[41]

phenomenon was paralleled by the HDAC-mediated downregulation of miR-328. Restoration of miR-328 revealed that miR-328 functioned as an decoy by binding hnRNP E2 to relapse CEBPA and promote myeloid differentiation [41]. *This was the first description of miRNA functioning as molecular decoys to suppress tumorigenesis.* In addition miR-328 also targeted the Pim1 kinase in CML, loss of miR-328 led to higher expression of Pim1 and CML survival in clonogenic assays [41]. The expression and physiological significance of miRNA important in CML are summarized in Table 4.

5 Conclusions

In summary, great strides have been made in indentifying miRNA profiles in every subtype of leukemia. These advances have already made assaying miRNA expression an accurate and sensitive tool in diagnosing leukemioa. Recent advances in the field have also provided major insights into the critical role that specific miRNA play in the genesis, pathobiology or chemo-resistance of leukemias. However, the cellular targets and function of the vast majority of miRNA that are deregulated in leukemia remain to be elucidated. In addition, not much is known about the mechanisms by which miRNA expression is altered in leukemia. Undoubtedly, the coming years will shed light on the myriad roles by which miRNA regulate cellular functions and will lead to the advent microRNAs as therapeutics in leukemias.

References

1. Agirre X, Vilas-Zornoza A, Jimenez-Velasco A, Martin-Subero JI, Cordeu L, Garate L, San Jose-Eneriz E, Abizanda G, Rodriguez-Otero P, Fortes P et al (2009) Epigenetic silencing of the tumor suppressor microRNA Hsa-miR-124a regulates CDK6 expression and confers a poor prognosis in acute lymphoblastic leukemia. Cancer Res 69:4443–4453
2. Asslaber D, Pinon JD, Seyfried I, Desch P, Stocher M, Tinhofer I, Egle A, Merkel O, Greil R (2010) MicroRNA-34a expression correlates with MDM2 SNP309 polymorphism and treatment-free survival in chronic lymphocytic leukemia. Blood 115:4191–4197
3. Baek D, Villen J, Shin C, Camargo FD, Gygi SP, Bartel DP (2008) The impact of microRNAs on protein output. Nature 455:64–71
4. Baldus CD, Martus P, Burmeister T, Schwartz S, Gokbuget N, Bloomfield CD, Hoelzer D, Thiel E, Hofmann WK (2007) Low ERG and BAALC expression identifies a new subgroup of adult acute T-lymphoblastic leukemia with a highly favorable outcome. J Clin Oncol 25:3739–3745
5. Bartel DP (2009) MicroRNAs: target recognition and regulatory functions. Cell 136:215–233
6. Bartel DP, Chen CZ (2004) Micromanagers of gene expression: the potentially widespread influence of metazoan microRNAs. Nat Rev Genet 5:396–400
7. Becker H, Marcucci G, Maharry K, Radmacher MD, Mrozek K, Margeson D, Whitman SP, Wu YZ, Schwind S, Paschka P et al (2010) Favorable prognostic impact of NPM1 mutations in older patients with cytogenetically normal de novo acute myeloid leukemia and associated gene- and microRNA-expression signatures: a Cancer and Leukemia Group B study. J Clin Oncol 28:596–604
8. Bhatia S, Kaul D, Varma N (2010) Potential tumor suppressive function of miR-196b in B-cell lineage acute lymphoblastic leukemia. Mol Cell Biochem 340:97–106
9. Bhatia S, Kaul D, Varma N (2011) Functional genomics of tumor suppressor miR-196b in T-cell acute lymphoblastic leukemia. Mol Cell Biochem 346:103–116
10. Blum W, Garzon R, Klisovic RB, Schwind S, Walker A, Geyer S, Liu S, Havelange V, Becker H, Schaaf L et al (2010) Clinical response and miR-29b predictive significance in older AML patients treated with a 10-day schedule of decitabine. Proc Natl Acad Sci U S A 107:7473–7478
11. Bonci D, Coppola V, Musumeci M, Addario A, Giuffrida R, Memeo L, D'Urso L, Pagliuca A, Biffoni M, Labbaye C et al (2008) The miR-15a-miR-16-1 cluster controls prostate cancer by targeting multiple oncogenic activities. Nat Med 14:1271–1277
12. Bousquet M, Harris MH, Zhou B, Lodish HF MicroRNA miR-125b causes leukemia. Proc Natl Acad Sci U S A 107:21558–21563
13. Brioschi M, Fischer J, Cairoli R, Rossetti S, Pezzetti L, Nichelatti M, Turrini M, Corlazzoli F, Scarpati B, Morra E et al (2010) Down-regulation of microRNAs 222/221 in acute myelogenous leukemia with deranged core-binding factor subunits. Neoplasia 12:866–876
14. Bueno MJ, Perez de Castro I, Gomez de Cedron M, Santos J, Calin GA, Cigudosa JC, Croce CM, Fernandez-Piqueras J, Malumbres M (2008) Genetic and epigenetic silencing of microRNA-203 enhances ABL1 and BCR-ABL1 oncogene expression. Cancer Cell 13:496–506
15. Calin GA, Croce CM (2006a) Genomics of chronic lymphocytic leukemia microRNAs as new players with clinical significance. Semin Oncol 33:167–173
16. Calin GA, Croce CM (2006b) MicroRNA signatures in human cancers. Nat Rev Cancer 6:857–866
17. Calin GA, Croce CM (2006c) MicroRNAs and chromosomal abnormalities in cancer cells. Oncogene 25:6202–6210
18. Calin GA, Croce CM (2007) Chromosomal rearrangements and microRNAs: a new cancer link with clinical implications. J Clin Invest 117:2059–2066
19. Calin GA, Dumitru CD, Shimizu M, Bichi R, Zupo S, Noch E, Aldler H, Rattan S, Keating M, Rai K et al (2002) Frequent deletions and down-regulation of micro- RNA genes miR15

and miR16 at 13q14 in chronic lymphocytic leukemia. Proc Natl Acad Sci U S A 99:15524–15529
20. Calin GA, Liu CG, Sevignani C, Ferracin M, Felli N, Dumitru CD, Shimizu M, Cimmino A, Zupo S, Dono M et al (2004a) MicroRNA profiling reveals distinct signatures in B cell chronic lymphocytic leukemias. Proc Natl Acad Sci U S A 101:11755–11760
21. Calin GA, Sevignani C, Dumitru CD, Hyslop T, Noch E, Yendamuri S, Shimizu M, Rattan S, Bullrich F, Negrini M et al (2004b) Human microRNA genes are frequently located at fragile sites and genomic regions involved in cancers. Proc Natl Acad Sci U S A 101:2999–3004
22. Calin GA, Ferracin M, Cimmino A, Di Leva G, Shimizu M, Wojcik SE, Iorio MV, Visone R, Sever NI, Fabbri M et al (2005) A MicroRNA signature associated with prognosis and progression in chronic lymphocytic leukemia. N Engl J Med 353:1793–1801
23. Calin GA, Cimmino A, Fabbri M, Ferracin M, Wojcik SE, Shimizu M, Taccioli C, Zanesi N, Garzon R, Aqeilan RI et al (2008) MiR-15a and miR-16–1 cluster functions in human leukemia. Proc Natl Acad Sci U S A 105:5166–5171
24. Cammarata G, Augugliaro L, Salemi D, Agueli C, La Rosa M, Dagnino L, Civiletto G, Messana F, Marfia A, Bica MG et al (2010) Differential expression of specific microRNA and their targets in acute myeloid leukemia. Am J Hematol 85:331–339
25. Carella C, Bonten J, Sirma S, Kranenburg TA, Terranova S, Klein-Geltink R, Shurtleff S, Downing JR, Zwarthoff EC, Liu PP et al (2007) MN1 overexpression is an important step in the development of inv(16) AML. Leukemia 21:1679–1690
26. Chandra P, Luthra R, Zuo Z, Yao H, Ravandi F, Reddy N, Garcia-Manero G, Kantarjian H, Jones D (2010) Acute myeloid leukemia with t(9;11)(p21–22;q23): common properties of dysregulated ras pathway signaling and genomic progression characterize de novo and therapy-related cases. Am J Clin Pathol 133:686–693
27. Chang TC, Wentzel EA, Kent OA, Ramachandran K, Mullendore M, Lee KH, Feldmann G, Yamakuchi M, Ferlito M, Lowenstein CJ et al (2007) Transactivation of miR-34a by p53 broadly influences gene expression and promotes apoptosis. Mol Cell 26:745–752
28. Chapiro E, Russell LJ, Struski S, Cave H, Radford-Weiss I, Valle VD, Lachenaud J, Brousset P, Bernard OA, Harrison CJ et al (2010) A new recurrent translocation t(11;14)(q24;q32) involving IGH@ and miR-125b-1 in B-cell progenitor acute lymphoblastic leukemia. Leukemia 24:1362–1364
29. Chase A, Huntly BJ, Cross NC (2001) Cytogenetics of chronic myeloid leukaemia. Best Pract Res Clin Haematol 14:553–571
30. Chim CS, Wong AS Kwong YL (2005) Epigenetic inactivation of the CIP/KIP cell-cycle control pathway in acute leukemias. Am J Hematol 80:282–287
31. Chim CS, Wong KY, Qi Y, Loong F, Lam WL, Wong LG, Jin DY, Costello JF, Liang R (2010) Epigenetic inactivation of the miR-34a in hematological malignancies. Carcinogenesis 31:745–750
32. Chim CS, Wong KY, Leung CY, Chung LP, Hui PK, Chan SY, Yu L (2011a) Epigenetic inactivation of the hsa-miR-203 in haematological malignancies. J Cell Mol Med 15(12):2760–2767
33. Chim CS, Wong KY, Qi Y, Loong F, Lam WL, Wong LG, Jin DY, Costello JF, Liang R (2011b) Epigenetic inactivation of the miR-34a in hematological malignancies. Carcinogenesis 31:745–750
34. Chuang JC, Jones PA (2007) Epigenetics and microRNAs. Pediatr Res 61:24R-29R
35. Cimmino A, Calin GA, Fabbri M, Iorio MV, Ferracin M, Shimizu M, Wojcik SE, Aqeilan RI, Zupo S, Dono M et al (2005) miR-15 and miR-16 induce apoptosis by targeting BCL2. Proc Natl Acad Sci U S A 102:13944–13949
36. Coskun E, von der Heide EK, Schlee C, Kuhnl A, Gokbuget N, Hoelzer D, Hofmann WK, Thiel E, Baldus CD (2011) The role of microRNA-196a and microRNA-196b as ERG regulators in acute myeloid leukemia and acute T-lymphoblastic leukemia. Leuk Res 35:208–213
37. Croce CM (2009) Causes and consequences of microRNA dysregulation in cancer. Nat Rev Genet 10:704–714

38. Davidsson J, Lilljebjorn H, Andersson A, Veerla S, Heldrup J, Behrendtz M, Fioretos T, Johansson B (2009) The DNA methylome of pediatric acute lymphoblastic leukemia. Hum Mol Genet 18:4054–4065
39. Dixon-McIver A, East P, Mein CA, Cazier JB, Molloy G, Chaplin T, Andrew Lister T, Young BD, Debernardi S (2008) Distinctive patterns of microRNA expression associated with karyotype in acute myeloid leukaemia. PLoS One 3:e2141
40. Ebert BL (2010) Genetic deletions in AML and MDS. Best Pract Res Clin Haematol 23:457–461
41. Eiring AM, Harb JG, Neviani P, Garton C, Oaks JJ, Spizzo R, Liu S, Schwind S, Santhanam R, Hickey CJ et al (2010) miR-328 functions as an RNA decoy to modulate hnRNP E2 regulation of mRNA translation in leukemic blasts. Cell 140:652–665
42. Eyholzer M, Schmid S, Wilkens L, Mueller BU, Pabst T (2010) The tumour-suppressive miR-29a/b1 cluster is regulated by CEBPA and blocked in human AML. Br J Cancer 103:275–284
43. Fabbri M, Bottoni A, Shimizu M, Spizzo R, Nicoloso MS, Rossi S, Barbarotto E, Cimmino A, Adair B, Wojcik SE et al (2011) Association of a microRNA/TP53 feedback circuitry with pathogenesis and outcome of B-cell chronic lymphocytic leukemia. JAMA 305:59–67
44. Farh KK, Grimson A, Jan C, Lewis BP, Johnston WK, Lim LP, Burge CB, Bartel DP (2005) The widespread impact of mammalian MicroRNAs on mRNA repression and evolution. Science 310:1817–1821
45. Fatica A, Rosa A, Fazi F, Ballarino M, Morlando M, De Angelis FG, Caffarelli E, Nervi C, Bozzoni I (2006) MicroRNAs and hematopoietic differentiation. Cold Spring Harb Symp Quant Biol 71:205–210
46. Fazi F, Racanicchi S, Zardo G, Starnes LM, Mancini M, Travaglini L, Diverio D, Ammatuna E, Cimino G, Lo-Coco F et al (2007) Epigenetic silencing of the myelopoiesis regulator microRNA-223 by the AML1/ETO oncoprotein. Cancer Cell 12:457–466
47. Ferracin M, Zagatti B, Rizzotto L, Cavazzini F, Veronese A, Ciccone M, Saccenti E, Lupini L, Grilli A, De Angeli C et al (2010) MicroRNAs involvement in fludarabine refractory chronic lymphocytic leukemia. Mol Cancer 9:123
48. Fulci V, Chiaretti S, Goldoni M, Azzalin G, Carucci N, Tavolaro S, Castellano L, Magrelli A, Citarella F, Messina M et al (2007) Quantitative technologies establish a novel microRNA profile of chronic lymphocytic leukemia. Blood 109:4944–4951
49. Fulci V, Colombo T, Chiaretti S, Messina M, Citarella F, Tavolaro S, Guarini A, Foa R, Macino G (2009) Characterization of B- and T-lineage acute lymphoblastic leukemia by integrated analysis of MicroRNA and mRNA expression profiles. Genes Chromosomes Cancer 48:1069–1082
50. Garzon R, Garofalo M, Martelli MP, Briesewitz R, Wang L, Fernandez-Cymering C, Volinia S, Liu CG, Schnittger S, Haferlach T et al (2008a) Distinctive microRNA signature of acute myeloid leukemia bearing cytoplasmic mutated nucleophosmin. Proc Natl Acad Sci U S A 105:3945–3950
51. Garzon R, Volinia S, Liu CG, Fernandez-Cymering C, Palumbo T, Pichiorri F, Fabbri M, Coombes K, Alder H, Nakamura T et al (2008b) MicroRNA signatures associated with cytogenetics and prognosis in acute myeloid leukemia. Blood 111:3183–3189
52. Garzon R, Heaphy CE, Havelange V, Fabbri M, Volinia S, Tsao T, Zanesi N, Kornblau SM, Marcucci G, Calin GA et al (2009a) MicroRNA 29b functions in acute myeloid leukemia. Blood 114:5331–5341
53. Garzon R, Heaphy CE, Havelange V, Fabbri M, Volinia S, Tsao T, Zanesi N, Kornblau SM, Marcucci G, Calin GA et al (2009b) MicroRNA 29b functions in acute myeloid leukemia. Blood 114(26):5331–5341
54. Garzon R, Liu S, Fabbri M, Liu Z, Heaphy CE, Callegari E, Schwind S, Pang J, Yu J, Muthusamy N et al (2009c) MicroRNA-29b induces global DNA hypomethylation and tumor suppressor gene reexpression in acute myeloid leukemia by targeting directly DNMT3A and 3B and indirectly DNMT1. Blood 113:6411–6418
55. Gefen N, Binder V, Zaliova M, Linka Y, Morrow M, Novosel A, Edry L, Hertzberg L, Shomron N, Williams O et al (2010) Hsa-mir-125b-2 is highly expressed in childhood ETV6/

RUNX1 (TEL/AML1) leukemias and confers survival advantage to growth inhibitory signals independent of p53. Leukemia 24:89–96
56. Grosveld GC (2007) MN1, a novel player in human AML. Blood Cells Mol Dis 39:336–339
57. Hamblin TJ (2007) Prognostic markers in chronic lymphocytic leukaemia. Best Pract Res Clin Haematol 20:455–468
58. Harrison CJ (2000) The genetics of childhood acute lymphoblastic leukaemia. Baillieres Best Pract Res Clin Haematol 13:427–439
59. Herman JG, Jen J, Merlo A, Baylin SB (1996) Hypermethylation-associated inactivation indicates a tumor suppressor role for p15INK4B. Cancer Res 56:722–727
60. Jima DD, Zhang J, Jacobs C, Richards KL, Dunphy CH, Choi WW, Yan Au W, Srivastava G, Czader MB, Rizzieri DA et al (2010) Deep sequencing of the small RNA transcriptome of normal and malignant human B cells identifies hundreds of novel microRNAs. Blood 116, e118–127
61. Jongen-Lavrencic M, Sun SM, Dijkstra MK, Valk PJ, Lowenberg B (2008) MicroRNA expression profiling in relation to the genetic heterogeneity of acute myeloid leukemia. Blood 111:5078–5085
62. Ju X, Li D, Shi Q, Hou H, Sun N, Shen B (2009) Differential microRNA expression in childhood B-cell precursor acute lymphoblastic leukemia. Pediatr Hematol Oncol 26:1–10
63. Kawano S, Miller CW, Gombart AF, Bartram CR, Matsuo Y, Asou H, Sakashita A, Said J, Tatsumi E, Koeffler HP (1999) Loss of p73 gene expression in leukemias/lymphomas due to hypermethylation. Blood 94:1113–1120
64. Keating MJ, Chiorazzi N, Messmer B, Damle RN, Allen SL, Rai KR, Ferrarini M, Kipps TJ (2003) Biology and treatment of chronic lymphocytic leukemia. Hematology, American Society of Hematology Education Program Book 2003(1):153–175
65. Klein U, Lia M, Crespo M, Siegel R, Shen Q, Mo T, Ambesi-Impiombato A, Califano A, Migliazza A, Bhagat G et al (2010) The DLEU2/miR-15a/16-1 Cluster Controls B Cell Proliferation and Its Deletion Leads to Chronic Lymphocytic Leukemia. Cancer Cell 17(1):28-40
66. Langer C, Marcucci G, Holland KB, Radmacher MD, Maharry K, Paschka P, Whitman SP, Mrozek K, Baldus CD, Vij R et al (2009) Prognostic importance of MN1 transcript levels, and biologic insights from MN1-associated gene and microRNA expression signatures in cytogenetically normal acute myeloid leukemia: a cancer and leukemia group B study. J Clin Oncol 27:3198–3204
67. Langer C, Radmacher MD, Ruppert AS, Whitman SP, Paschka P, Mrozek K, Baldus CD, Vukosavljevic T, Liu CG, Ross ME et al (2008) High BAALC expression associates with other molecular prognostic markers, poor outcome, and a distinct gene-expression signature in cytogenetically normal patients younger than 60 years with acute myeloid leukemia: a Cancer and Leukemia Group B (CALGB) study. Blood 111:5371–5379
68. Lerner M, Harada M, Loven J, Castro J, Davis Z, Oscier D, Henriksson M, Sangfelt O, Grander D, Corcoran MM (2009) DLEU2, frequently deleted in malignancy, functions as a critical host gene of the cell cycle inhibitory microRNAs miR-15a and miR-16-1. Exp Cell Res 315:2941–2952
69. Levitt L, Lin R (1996) Biology and treatment of adult acute lymphoblastic leukemia. West J Med 164:143–155
70. Li S, Moffett HF, Lu J, Werner L, Zhang H, Ritz J, Neuberg D, Wucherpfennig KW, Brown JR, Novina CD (2010a) MicroRNA expression profiling identifies activated B cell status in chronic lymphocytic leukemia cells. PLoS One 6:e16956
71. Li S, Moffett HF, Lu J, Werner L, Zhang H, Ritz J, Neuberg D, Wucherpfennig KW, Brown JR, Novina CD (2011) MicroRNA expression profiling identifies activated B cell status in chronic lymphocytic leukemia cells. PLoS One 6:e16956
72. Li X, Liu J, Zhou R, Huang S, Chen XM (2010b) Gene silencing of MIR22 in acute lymphoblastic leukaemia involves histone modifications independent of promoter DNA methylation. Br J Haematol 148:69–79

73. Li Z, Lu J, Sun M, Mi S, Zhang H, Luo RT, Chen P, Wang Y, Yan M, Qian Z et al (2008) Distinct microRNA expression profiles in acute myeloid leukemia with common translocations. Proc Natl Acad Sci U S A 105:15535–15540
74. Liu M, Taketani T, Li R, Takita J, Taki T, Yang HW, Kawaguchi H, Ida K, Matsuo Y, Hayashi Y (2001) Loss of p73 gene expression in lymphoid leukemia cell lines is associated with hypermethylation. Leuk Res 25:441–447
75. Liu S, Wu LC, Pang J, Santhanam R, Schwind S, Wu YZ, Hickey CJ, Yu J, Becker H, Maharry K et al (2010) Sp1/NFkappaB/HDAC/miR-29b regulatory network in KIT-driven myeloid leukemia. Cancer Cell 17:333–347
76. Lu J, Getz G, Miska EA, Alvarez-Saavedra E, Lamb J, Peck D, Sweet-Cordero A, Ebert BL, Mak RH, Ferrando AA et al (2005) MicroRNA expression profiles classify human cancers. Nature 435:834–838
77. Machova Polakova K, Lopotova T, Klamova H, Burda P, Trneny M, Stopka T, Moravcova J (2011) Expression patterns of microRNAs associated with CML phases and their disease related targets. Mol Cancer 10:41
78. Marcucci G, Maharry K, Radmacher MD, Mrozek K, Vukosavljevic T, Paschka P, Whitman SP, Langer C, Baldus CD, Liu, CG et al (2008a) Prognostic significance of, and gene and microRNA expression signatures associated with, CEBPA mutations in cytogenetically normal acute myeloid leukemia with high-risk molecular features: a Cancer and Leukemia Group B Study. J Clin Oncol 26:5078–5087
79. Marcucci G, Radmacher MD, Maharry K, Mrozek K, Ruppert AS, Paschka P, Vukosavljevic T, Whitman SP, Baldus CD, Langer C et al (2008b) MicroRNA expression in cytogenetically normal acute myeloid leukemia. N Engl J Med 358:1919–1928
80. Marcucci G, Radmacher MD, Mrozek K, Bloomfield, CD (2009) MicroRNA expression in acute myeloid leukemia. Curr Hematol Malig Rep 4:83–88
81. Marcucci G, Haferlach T, Dohner, H (2011) Molecular genetics of adult acute myeloid leukemia: prognostic and therapeutic implications. J Clin Oncol 29:475–486
82. Marton S, Garcia MR, Robello C, Persson H, Trajtenberg F, Pritsch O, Rovira C, Naya H, Dighiero G, Cayota, A (2008) Small RNAs analysis in CLL reveals a deregulation of miRNA expression and novel miRNA candidates of putative relevance in CLL pathogenesis. Leukemia 22:330–338
83. Matsushita C, Yang Y, Takeuchi S, Matsushita M, Van Dongen JJ, Szczepanski T, Bartram CR, Seo H, Koeffler HP, Taguchi, H (2004) Aberrant methylation in promoter-associated CpG islands of multiple genes in relapsed childhood acute lymphoblastic leukemia. Oncol Rep 12:97–99
84. Mavrakis KJ, Wolfe AL, Oricchio E, Palomero T, de Keersmaecker K, McJunkin K, Zuber J, James T, Khan AA, Leslie, CS et al (2011) Genome-wide RNA-mediated interference screen identifies miR-19 targets in Notch-induced T-cell acute lymphoblastic leukaemia. Nat Cell Biol 12:372–379
85. Mehes, G (2005) Chromosome abnormalities with prognostic impact in B-cell chronic lymphocytic leukemia. Pathol Oncol Res 11:205–210
86. Mertens D, Philippen A, Ruppel M, Allegra D, Bhattacharya N, Tschuch C, Wolf S, Idler I, Zenz T, Stilgenbauer, S (2009) Chronic lymphocytic leukemia and 13q14: miRs and more. Leuk Lymphoma 50:502–505
87. Mi S, Lu J, Sun M, Li Z, Zhang H, Neilly MB, Wang Y, Qian Z, Jin J, Zhang Y et al (2007) MicroRNA expression signatures accurately discriminate acute lymphoblastic leukemia from acute myeloid leukemia. Proc Natl Acad Sci U S A 104:19971–19976
88. Moussay E, Wang K, Cho JH, van Moer K, Pierson S, Paggetti J, Nazarov PV, Palissot V, Hood LE, Berchem G et al (2011) MicroRNA as biomarkers and regulators in B-cell chronic lymphocytic leukemia. Proc Natl Acad Sci U S A 108:6573–6578
89. Mraz M, Malinova K, Kotaskova J, Pavlova S, Tichy B, Malcikova J, Stano Kozubik K, Smardova J, Brychtova Y, Doubek M et al (2009a) miR-34a, miR-29c and miR-17-5p are downregulated in CLL patients with TP53 abnormalities. Leukemia 23:1159–1163

90. Mraz M, Pospisilova S, Malinova K, Slapak I, Mayer, J (2009b) MicroRNAs in chronic lymphocytic leukemia pathogenesis and disease subtypes. Leuk Lymphoma 50:506–509
91. Mrozek, K (2008) Cytogenetic, molecular genetic, and clinical characteristics of acute myeloid leukemia with a complex karyotype. Semin Oncol 35:365–377
92. Mrozek K, Bloomfield, CD (2006) Chromosome aberrations, gene mutations and expression changes, and prognosis in adult acute myeloid leukemia. Hematology, American Society of Hematology Education Program Book 2006(1):169–177
93. Mrozek K, Bloomfield, CD (2008) Clinical significance of the most common chromosome translocations in adult acute myeloid leukemia. J Natl Cancer Inst Monogr (39):52–57
94. Mrozek K, Heerema NA, Bloomfield, CD (2004) Cytogenetics in acute leukemia. Blood Rev 18:115–136
95. Nagarajan, L (2010) Chromosomal deletions in AML. Cancer Treat Res 145:59–66
96. Nagel S, Venturini L, Przybylski GK, Grabarczyk P, Schmidt CA, Meyer C, Drexler HG, Macleod RA, Scherr, M (2009) Activation of miR-17-92 by NK-like homeodomain proteins suppresses apoptosis via reduction of E2F1 in T-cell acute lymphoblastic leukemia. Leuk Lymphoma 50:101–108
97. Ouillette P, Erba H, Kujawski L, Kaminski M, Shedden K, Malek, SN (2008) Integrated genomic profiling of chronic lymphocytic leukemia identifies subtypes of deletion 13q14. Cancer Res 68:1012–1021
98. Pabst T, Mueller BU, Harakawa N, Schoch C, Haferlach T, Behre G, Hiddemann W, Zhang DE, Tenen, DG (2001a) AML1-ETO downregulates the granulocytic differentiation factor C/EBPalpha in t(8;21) myeloid leukemia. Nat Med 7:444–451
99. Pabst T, Mueller BU, Zhang P, Radomska HS, Narravula S, Schnittger S, Behre G, Hiddemann W, Tenen, DG (2001b) Dominant-negative mutations of CEBPA, encoding CCAAT/enhancer binding protein-alpha (C/EBPalpha), in acute myeloid leukemia. Nat Genet 27:263–270
100. Pallasch CP, Patz M, Park YJ, Hagist S, Eggle D, Claus R, Debey-Pascher S, Schulz A, Frenzel LP, Claasen J et al (2009) miRNA deregulation by epigenetic silencing disrupts suppression of the oncogene PLAG1 in chronic lymphocytic leukemia. Blood 114:3255–3264
101. Pekarsky Y, Santanam U, Cimmino A, Palamarchuk A, Efanov A, Maximov V, Volinia S, Alder H, Liu CG, Rassenti L et al (2006) Tcl1 expression in chronic lymphocytic leukemia is regulated by miR-29 and miR-181. Cancer Res 66:11590–11593
102. Petrocca F, Vecchione A, Croce, CM (2008a) Emerging role of miR-106b-25/miR-17-92 clusters in the control of transforming growth factor beta signaling. Cancer Res 68:8191–8194
103. Petrocca F, Visone R, Onelli MR, Shah MH, Nicoloso MS, de Martino I, Iliopoulos D, Pilozzi E, Liu CG, Negrini M et al (2008b) E2F1-regulated microRNAs impair TGFbeta-dependent cell-cycle arrest and apoptosis in gastric cancer. Cancer Cell 13:272–286
104. Pulikkan JA, Peramangalam PS, Dengler V, Ho PA, Preudhomme C, Meshinchi S, Christopeit M, Nibourel O, Muller-Tidow C, Bohlander, SK et al (2010) C/EBPalpha regulated microRNA-34a targets E2F3 during granulopoiesis and is down-regulated in AML with CEBPA mutations. Blood 116:5638–5649
105. Rassenti LZ, Huynh L, Toy TL, Chen L, Keating MJ, Gribben JG, Neuberg DS, Flinn IW, Rai KR, Byrd, JC et al (2004) ZAP-70 compared with immunoglobulin heavy-chain gene mutation status as a predictor of disease progression in chronic lymphocytic leukemia. N Engl J Med 351:893–901
106. Raveche ES, Salerno E, Scaglione BJ, Manohar V, Abbasi F, Lin YC, Fredrickson T, Landgraf P, Ramachandra S, Huppi K et al (2007) Abnormal microRNA-16 locus with synteny to human 13q14 linked to CLL in NZB mice. Blood 109:5079–5086
107. Raver-Shapira N, Marciano E, Meiri E, Spector Y, Rosenfeld N, Moskovits N, Bentwich Z, Oren M (2007) Transcriptional activation of miR-34a contributes to p53-mediated apoptosis. Mol Cell 26:731–743

108. Reuss-Borst MA, Buhring HJ, Schmidt H, Muller, CA (1994) AML: immunophenotypic heterogeneity and prognostic significance of c-kit expression. Leukemia 8:258–263
109. Roman-Gomez J, Jimenez-Velasco A, Barrios M, Prosper F, Heiniger A, Torres A, Agirre, X (2007) Poor prognosis in acute lymphoblastic leukemia may relate to promoter hypermethylation of cancer-related genes. Leuk Lymphoma 48:1269–1282
110. Roman-Gomez J, Agirre X, Jiménez-Velasco A, Arqueros V, Vilas-Zornoza A, Rodriguez-Otero P, Martin-Subero I, Garate L, Cordeu L, San José-Eneriz E, Martin V, Castillejo JA, Bandrés E, Calasanz MJ, Siebert R, Heiniger A, Torres A, Prosper F (2009). Epigenetic regulation of microRNAs in acute lymphoblastic leukemia. J Clin Oncol 27(8):1316–1322
111. Rossi S, Shimizu M, Barbarotto E, Nicoloso MS, Dimitri F, Sampath D, Fabbri M, Lerner S, Barron LL, Rassenti, LZ et al (2010) microRNA fingerprinting of CLL patients with chromosome 17p deletion identify a miR-21 score that stratifies early survival. Blood 116(6):945–952
112. Sampath D, Calin GA, Puduvalli VK, Gopisetty G, Taccioli C, Liu CG, Ewald B, Liu C, Keating MJ, Plunkett, W (2009) Specific activation of microRNA106b enables the p73 apoptotic response in chronic lymphocytic leukemia by targeting the ubiquitin ligase Itch for degradation. Blood 113:3744–3753
113. San Jose-Eneriz E, Roman-Gomez J, Jimenez-Velasco A, Garate L, Martin V, Cordeu L, Vilas-Zornoza A, Rodriguez-Otero P, Calasanz MJ, Prosper F et al (2009) MicroRNA expression profiling in Imatinib-resistant Chronic Myeloid Leukemia patients without clinically significant ABL1-mutations. Mol Cancer 8:69
114. Santanam U, Zanesi N, Efanov A, Costinean S, Palamarchuk A, Hagan JP, Volinia S, Alder H, Rassenti L, Kipps T et al (2010) Chronic lymphocytic leukemia modeled in mouse by targeted miR-29 expression. Proc Natl Acad Sci U S A 107:12210–12215
115. Schotte D, Chau JC, Sylvester G, Liu G, Chen C, Van Der Velden VH, Broekhuis MJ, Peters TC, Pieters R, den Boer, ML (2009) Identification of new microRNA genes and aberrant microRNA profiles in childhood acute lymphoblastic leukemia. Leukemia 23:313–322
116. Schotte D, De Menezes RX, Moqadam FA, Khankahdani LM, Lange-Turenhout E, Chen C, Pieters R, Den Boer, ML (2011) MicroRNA characterize genetic diversity and drug resistance in pediatric acute lymphoblastic leukemia. Haematologica 96:703–711
117. Schotte D, Moqadam FA, Lange-Turenhout EA, Chen C, van Ijcken WF, Pieters R, den Boer, ML (2011b) Discovery of new microRNAs by small RNAome deep sequencing in childhood acute lymphoblastic leukemia. Leukemia 25(9):1389-1399
118. Schwind S, Maharry K, Radmacher MD, Mrozek K, Holland KB, Margeson D, Whitman SP, Hickey C, Becker H, Metzeler, KH et al (2010a) Prognostic significance of expression of a single microRNA, miR-181a, in cytogenetically normal acute myeloid leukemia: a Cancer and Leukemia Group B study. J Clin Oncol 28:5257–5264
119. Schwind S, Marcucci G, Maharry K, Radmacher MD, Mrozek K, Holland KB, Margeson D, Becker H, Whitman SP, Wu YZ et al (2010b) BAALC and ERG expression levels are associated with outcome and distinct gene and microRNA expression profiles in older patients with de novo cytogenetically normal acute myeloid leukemia: a Cancer and Leukemia Group B study. Blood 116:5660–5669
120. Shin C, Nam JW, Farh KK, Chiang HR, Shkumatava A, Bartel, DP (2010) Expanding the microRNA targeting code: functional sites with centered pairing. Mol Cell 38:789–802
121. Stamatopoulos B, Meuleman N, Haibe-Kains B, Saussoy P, Van Den Neste E, Michaux L, Heimann P, Martiat P, Bron D, Lagneaux, L (2009) microRNA-29c and microRNA-223 down-regulation has *in vivo* significance in chronic lymphocytic leukemia and improves disease risk stratification. Blood 113:5237–5245
122. Stumpel DJ, Schotte D, Lange-Turenhout EA, Schneider P, Seslija L, de Menezes RX, Marquez VE, Pieters R, den Boer ML, Stam, RW (2011) Hypermethylation of specific microRNA genes in MLL-rearranged infant acute lymphoblastic leukemia: major matters at a micro scale. Leukemia 25:429–439

123. Suresh S, McCallum L, Lu W, Lazar N, Perbal B, Irvine, AE (2011) MicroRNAs 130a/b are regulated by BCR-ABL and downregulate expression of CCN3 in CML. J Cell Commun Signal
124. Visone R, Rassenti LZ, Veronese A, Taccioli C, Costinean S, Aguda BD, Volinia S, Ferracin M, Palatini J, Balatti V et al (2009) Karyotype-specific microRNA signature in chronic lymphocytic leukemia. Blood 114:3872–3879
125. Visone R, Veronese A, Rassenti LZ, Balatti V, Pearl DK, Acunzo M, Volinia S, Taccioli C, Kipps TJ, Croce, CM (2011) MiR-181b is a biomarker of disease progression in chronic lymphocytic leukemia. Blood
126. Wattanawaraporn R, Singhsilarak T, Nuchprayoon I, Mutirangura, A (2007) Hypermethylation of TTC12 gene in acute lymphoblastic leukemia. Leukemia 21:2370–2373
127. Whitman SP, Maharry K, Radmacher MD, Becker H, Mrozek K, Margeson D, Holland KB, Wu YZ, Schwind S, Metzeler, KH et al (2010) FLT3 internal tandem duplication associates with adverse outcome and gene- and microRNA-expression signatures in patients 60 years of age or older with primary cytogenetically normal acute myeloid leukemia: a Cancer and Leukemia Group B study. Blood 116:3622–3626
128. Zanette DL, Rivadavia F, Molfetta GA, Barbuzano FG, Proto-Siqueira R, Silva-Jr WA, Falcao RP, Zago, MA (2007) miRNA expression profiles in chronic lymphocytic and acute lymphocytic leukemia. Braz J Med Biol Res 40:1435–1440
129. Zenz T, Benner A, Dohner H, Stilgenbauer, S (2008) Chronic lymphocytic leukemia and treatment resistance in cancer: the role of the p53 pathway. Cell Cycle 7:3810–3814
130. Zenz T, Habe S, Denzel T, Mohr J, Winkler D, Buhler A, Sarno A, Groner S, Mertens D, Busch R et al (2009a) Detailed analysis of p53 pathway defects in fludarabine-refractory chronic lymphocytic leukemia (CLL): dissecting the contribution of 17p deletion, TP53 mutation, p53-p21 dysfunction, and miR34a in a prospective clinical trial. Blood 114:2589–2597
131. Zenz T, Mohr J, Eldering E, Kater AP, Buhler A, Kienle D, Winkler D, Durig J, van Oers MH, Mertens D et al (2009b) miR-34a as part of the resistance network in chronic lymphocytic leukemia. Blood 113:3801–3808
132. Zhang H, Yang JH, Zheng YS, Zhang P, Chen X, Wu J, Xu L, Luo XQ, Ke ZY, Zhou H et al (2009) Genome-wide analysis of small RNA and novel MicroRNA discovery in human acute lymphoblastic leukemia based on extensive sequencing approach. PLoS One 4:e6849
133. Zhao JJ, Lin J, Lwin T, Yang H, Guo J, Kong W, Dessureault S, Moscinski LC, Rezania D, Dalton WS et al (2010) microRNA expression profile and identification of miR-29 as a prognostic marker and pathogenetic factor by targeting CDK6 in mantle cell lymphoma. Blood 115:2630–2639
134. Zimmerman EI, Dollins CM, Crawford M, Grant S, Nana-Sinkam SP, Richards KL, Hammond SM, Graves LM (2010) Lyn kinase-dependent regulation of miR181 and myeloid cell leukemia-1 expression: implications for drug resistance in myelogenous leukemia. Mol Pharmacol 78:811–817

Small-Molecule Regulation of MicroRNA Function

Colleen M. Connelly and Alexander Deiters

Abstract MicroRNAs (miRNAs) are single-stranded noncoding RNAs of 21–23 nucleotides, which regulate the expression of genes by binding to the 3′ untranslated regions of target messenger RNAs (mRNAs). MicroRNAs down-regulate gene expression by either inhibiting translation or accelerating the degradation of the mRNA. It is estimated that miRNAs are involved in the regulation of about 30 % of all genes and almost every genetic pathway, making miRNAs an important class of gene regulators. Variations in miRNA expression are involved in many human diseases including cancer, immune disorders, diabetes, and cardiovascular diseases. Thus, small molecule modifiers of miRNA function have potential as new therapeutic agents, as probes for the elucidation of detailed mechanisms of miRNA function and regulation, and as tools for the discovery of new targets for the treatment of human diseases. A variety of different assay systems have been developed and used in the discovery of small molecule modifiers of miRNA function. Identified small molecules regulate the miRNA pathway in either a general or a miRNA-specific fashion. The discovery and development of these molecules demonstrates that the miRNA pathway represents a feasible small molecule target. Several of these small molecules have also shown therapeutic potential in cell based experiments, supporting the idea that modifiers of miRNA function could lead to the identification of new drugs.

1 Introduction to MicroRNAs

MicroRNAs (miRNAs) were discovered in 1993 by Victor Ambros in the course of studies of the gene *lin-14* in the development of the nematode *C. elegans* [1]. These experiments identified two small RNA transcripts of 22 and 61 nucleotides that contain sequences complementary to regions of the 3′ untranslated region (UTR) of *lin-14*. Later experiments showed that the complementarities of the small RNAs

C. M. Connelly (✉) · A. Deiters
Department of Chemistry, North Carolina State University, Raleigh, NC 27695-8204, USA

A. Deiters
e-mail: alex_deiters@ncsu.edu

S. Alahari (ed.), *MicroRNA in Cancer*,
DOI 10.1007/978-94-007-4655-8_8, © Springer Science+Business Media Dordrecht 2013

to the *lin-14* mRNA was sufficient in inhibiting its translation to Lin-14 protein [2]. Another small RNA, named let-7, of 21 nucleotides was discovered in 2000 and showed complementarity to regions of the 3′ UTRs of the genes *lin-14*, *lin-28*, *lin-41*, *lin-42*, and *daf-12* [2]. Subsequent experiments showed the conservation of let-7 in a wide range of animal species including vertebrate, ascidian, hemichordate, mollusc, annelid, and arthropod, suggesting that miRNAs may represent a general class of gene regulators [3]. Since their discovery, many researchers have begun to focus on the identification, biogenesis, regulation, function, and therapeutic potential of miRNAs [4–8].

MicroRNAs are single stranded non-coding RNAs of 21–23 nucleotides and it is estimated that over 1,000 miRNAs exist in humans [9]. These endogenous miRNAs regulate gene expression in a sequence specific fashion by binding the 3′ untranslated region (UTR) of target messenger RNA (mRNA). The outcome of miRNA binding is the down regulation of the target mRNA through either suppression of translation or degradation [10]. However, most recently examples for 3′ UTR binding and the up-regulation of the expression of targeted genes have been reported as well [11, 12]. Most miRNAs target a multitude of different mRNA transcripts and it is estimated that miRNAs are involved in the regulation of up to 30 % of all genes and almost every genetic pathway. Therefore, misregulation of miRNA expression leads to the misregulation of its various mRNA targets and can have severe implications to the homeostasis of cells and tissues. miRNAs are involved in many biological processes including development, differentiation, apoptosis, and proliferation [13]. Therefore, the aberrant expression of certain miRNAs has been linked to a wide range of human diseases including cancer, immune disorders, diabetes, and cardiovascular diseases [14]. In particular, miRNA misregulation has been implicated in the initiation, progression, and metastasis of cancer and cellular resistance to apoptosis [9]. It has been observed that human cancers commonly display an altered expression profile of miRNAs with known anti-apoptotic (e.g., miR-21, miR-155, or miR-215) or tumor suppressive activity (e.g., let-7a, miR-15, miR-16, miR-34a, or miR-143/145). In fact, many different miRNAs have been linked to a variety of malignancies and it was recently reported that up to 192 miRNAs were abnormally expressed in cancer cells [15]. Given the fact that miRNAs play such a critical role in cancer and other diseases, it is important to understand the molecular details of their biogenesis, regulation, and function. Such information will be crucial in order to explore the causality of miRNA misregulation in human diseases, i.e., is the miRNA the disease causing factor or is its aberrant regulation a result of the disease? In one case, miR-21 (see Sect. 5.1), a clear correlation between cancer manifestation and miRNA overexpression was recently reported [16].

1.1 MicroRNA Biogenesis and Function

The majority of endogenous miRNAs are transcribed from the genome by RNA polymerase II into primary miRNAs (pri-miRNAs) which range from hundreds to

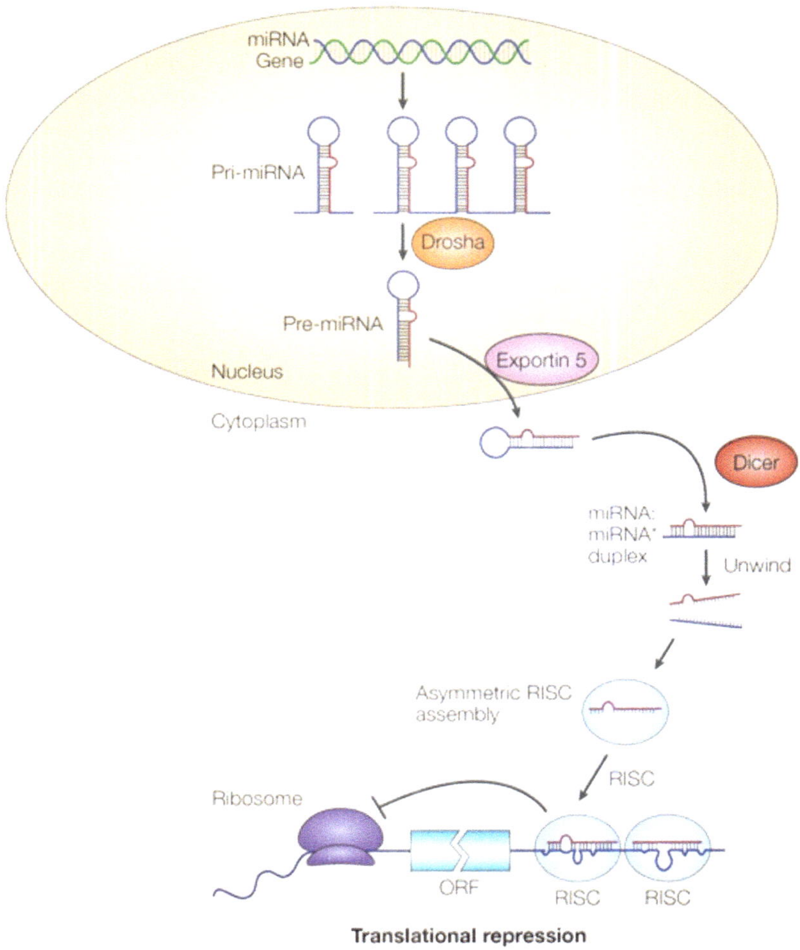

Fig. 1 The miRNA pathway depicting the transcription and processing of miRNAs and the miRNA induced translational repression of target mRNAs. (Adapted from [25])

thousands of nucleotides in length and contain at least one hairpin structure [17]. Pri-miRNAs are processed in the nucleus by the RNaseIII enzyme Drosha, with the aid of DGCR8, into shorter stem-loop structured double-stranded RNA molecules named precursor miRNAs or pre-miRNAs (Fig. 1). The cleavage of pri-miRNAs by Drosha generates a 5′ phosphate and a two nucleotide 3′ overhang on the pre-miRNA molecule [17]. Pre-miRNAs are approximately 65 nucleotides in length and contain the mature miRNA in either the 5′ or the 3′ half of the stem [18]. Pre-miRNAs are then transported from the nucleus to the cytoplasm by exportin 5, where they are further cleaved by the enzyme Dicer in concert with TAR RNA binding protein (TRBP) to produce the miRNA/miRNA* double-stranded RNA duplex

[19–23]. The miRNA/miRNA* duplex is unwound and one strand is preferentially loaded into the RNA-induced silencing complex (RISC) where the mature miRNA can target its single-stranded complementary mRNAs. The miRNAs bind to the 3′ untranslated region (UTR) of target mRNAs with imperfect complementarity; therefore, each miRNA can target several different mRNA transcripts [24]. This leads to the down regulation of gene expression through inhibition of translation by either blocking elongation initiation [25, 26] or sequestering the target mRNAs in P bodies away from the translational machinery [27]. Similarly, miRNA/RISC binding to target mRNAs can lead to mRNA cleavage and degradation, a mechanism that is more commonly seen in plants where miRNAs exhibit a higher degree of complementarity to their target mRNAs [28].

1.2 Endogenous MicroRNA Regulation

As previously discussed, mature miRNAs directly affect the translation of their corresponding mRNA targets; therefore, a misregulation in the expression of miRNAs leads to a misregulation of the target mRNAs. Because miRNAs are key regulators in cellular function, Nature has developed several regulatory mechanisms to control the expression level of individual miRNAs [29, 30]. The first type of regulation occurs pre-transcription. Regulation at this level can be affected by a change in miRNA gene copy number, a mutation in the miRNA gene, or histone deacetylation and hypermethylation of miRNA promoter regions [31–33]. For example, the tumor suppressive miRNA miR-127, which regulates the human proto-oncogene *BCL6*, is down-regulated or silenced in a variety of tumors. It has been shown that in the T24 bladder cancer cell line, the down-regulation of miR-127 is due to a high level of methylation in the promoter region of *miR-127* and that treatment of the T24 cells with demethylating agents leads to an increased expression of miR-127 [34]. As mentioned, pre-transcriptional regulation can also be affected by a miRNA gene mutation. It was reported that altered genomic loci containing miRNA genes occurred in 37 % of ovarian cancers, 73 % of breast cancers, and 86 % of melanoma cases [35].

Similarly, miRNA expression can be regulated at the transcriptional level. Transcription factors can directly regulate the expression of specific miRNAs and a number of transcription factors that regulate the levels of cancer-related miRNAs have been identified [36]. Many of these transcription factors bind to corresponding regulatory motifs in the upstream regions of miRNAs and allow for the recruitment of coactivators and the transcriptional machinery. The proto-oncogene *c-Myc* encodes the transcription factor c-Myc which activates the expression of a cluster of miRNAs miR-17-92 on human chromosome 13 [37]. c-Myc activates another transcription factor, E2F1 that promotes cell cycle progression. Interestingly, E2F1 is negatively regulated by two miRNAs of the miR-17-92 cluster, miR-17-5p and miR-20a. This system allows for the activation of the *E2F1* transcript by c-Myc and subsequent regulation of its translation by the miR-17-92 cluster. Another example of transcriptional regulation is found in the case of miRNA miR-133b, which regulates its own transcription through a negative feedback loop by targeting its own transcription factor in mammalian midbrain dopaminergic neurons [38]. The transcription of miR-133b is induced by the transcription factor PITX3. Subsequently, the expression of PITX3 is suppressed by the mature miR-133b.

Endogenous miRNA regulation can also occur post-transcriptionally. As discussed previously, miRNA processing is controlled by two main enzymes, Drosha and Dicer. Experiments have shown that the widespread down-regulation of miRNAs in tumors is caused by a failure at the Drosha processing step [39]. For example, studies have shown a post-transcriptional regulation in the processing of pri-miR-21 in human vascular smooth muscle cells controlled by transforming growth factor β (TGF-β) and bone morphogenetic protein (BMP) [40]. TGF-β and BMP specific SMAD signal transducers bind to the Drosha subunit p68, an RNA helicase, to recruit pri-miR-21. TGF-β and BMP signaling induces an increase in pri-miR-21 processing into pre-miR-21 and leads to an increase in mature miR-21 levels. In addition, it has been shown that Dicer accumulation is dependent on TRBP and that a decrease or mutation in TRBP leads to decreased Dicer stability and can result in general pre-miRNA processing defects [29].

At the post-transcriptional level, other proteins in addition to the general components of the miRNA pathway, may be involved in the processing of specific miRNAs or miRNA families. A recent discovery showed the KH-type splicing regulatory protein (KSRP), a mediator of mRNA decay, promotes the processing of several miRNAs by serving as a component of both the Drosha and Dicer complexes [41]. Transient knockdown of KSRP in HeLa cells led to more than 1.5-fold reduction in the levels of 14 miRNAs including let-7a, miR-26b, miR-20, miR-106a, miR-21, and miR-16 and a 1.2–1.5 fold reduction in expression of 20 more miRNAs. KSRP binds to the terminal loop of the target pre-miRNAs and promotes their recruitment and/or positioning in the processing complexes through protein-protein interactions [41]. The RNA binding protein LIN-28 was also shown to repress the processing of let-7 miRNAs by binding the terminal loop of pri-let-7 and interfering with Drosha cleavage in embryonic stem cells [42]. LIN-28 can also bind to pre-let-7 inducing a 3′ polyuridylation that prevents Dicer cleavage [43, 44].

1.3 Exogenous MicroRNA Regulation

Exogenous tools for miRNA regulation have been developed in order to investigate the function of miRNAs and to validate their cellular mRNA targets. Amongst these tools are miRNA antisense oligonucleotides (termed antagomirs), miRNA sponges, and miRNA over-expression vectors [45–47]. Antagomirs are chemically modified RNA oligonucleotides that are perfectly complementary to a miRNA of interest. Antagomirs act as competitive inhibitors of miRNA and target mRNA binding with complete specificity. Common modifications to make antagomirs more resistant to degradation include 2′ O-methylation (2′-OMe) of the nucleotides and the introduction of phosphorothioate bonds in the RNA backbone. The conjugation of a cholesterol moiety at the 3′ end of antagomirs has been utilized to improve the pharmacokinetic properties and enhance cellular uptake [48]. Another modification is the addition of a bridging methylene group between the 2′-O and the 4′-C of the ribose ring, generating a molecule known as a locked nucleic acid or LNA. LNAs have increased stability due to the bridging carbon. LNAs "lock" the ribose confirmation and make the molecule more resistant to exo- and endonucleases than other RNA based antisense oligonucleotides [49]. This modification also provides the oligonucleotides with superior miRNA affinity, mismatch discrimination, and low toxicity [49].

Other inhibitors of miRNA function include miRNA sponges or miRNA decoys. miRNA sponges are transcripts expressed from strong promoters that contain multiple, tandem binding sites for a miRNA of interest [50]. The transcripts act as competitive inhibitors to miRNA:mRNA binding. When vectors encoding these sponges are transfected into cells, the sponges act as a decoy to sequester the target miRNA [51]. In other words, the sponges are able to interact with the corresponding miRNA and prevent its association with its endogenous targets [48].

In contrast to the previously mentioned miRNA inhibitors, miRNA overexpression vectors increase miRNA levels, e.g., to restore the levels of underexpressed miRNAs. A potential gene therapy application of this approach is to raise the levels of miRNAs whose expression is downregulated in diseases such as cancer in order to restore the natural regulation of the miRNA target genes [48].

2 Introduction to Small Molecule Modifiers

In addition to the nucleic acid based tools for miRNA regulation, small molecule modifiers of miRNA function have been discovered in recent years and could provide both new probes for the investigation of miRNA biogenesis and function and potential therapeutics. Small molecules are low molecular weight organic compounds of typically <800 Da. They can bind to biological macromolecules, typically proteins, but also DNA and RNA, thus altering their activity or function. Small molecules can be of natural or synthetic origin, and can be optimized to have ideal drug properties including good efficacy, good solubility, good bioavailability, and good pharmacokinetics [52].

2.1 Small Molecule Modifiers of MicroRNA Function as Molecular Probes

Understanding the roles of miRNAs in cancer and other human malignancies has become increasingly important. Similar to the nucleic acid based tools that control miRNA function which were mentioned previously, small molecules can be used to explore the regulation and biogenesis of miRNAs. Small molecule modifiers of miRNAs can be used to study the response of miRNAs to compound treatment as well as the overall effect on the biological system. Small molecules have several advantages over nucleic acid based tools, making them a promising area of research. Because of their size, small molecules can diffuse across cell membranes more readily than macromolecules. Small molecules are also more stable intracellularly, because they are not susceptible to nucleases. An increased stability can be beneficial in cell culture environments and certainly for *in vivo* studies. Small molecules enable temporal control of miRNA function, since they can be added or withdrawn from the system under investigation at any given timepoint. Most importantly, small molecules have the potential to control miRNA expression on multiple levels of the miRNA pathway (Fig. 1) in contrast to antagomirs that can only regulate miRNA function through a direct interaction with the mature miRNA. Consequently, small molecules are more versatile probes to study miRNA biogenesis and regulation.

2.2 Small Molecule Modifiers of MicroRNA Function as Potential Therapeutics

Nucleic acid based molecules also face difficulties as therapeutics, such as low oral bioavailability, enzymatic degradation, poor cellular uptake, off-target effects, and activation of immune responses [7, 15]. Because of the role of miRNAs in cancer and other diseases, small molecule regulators of miRNA expression are being investigated for their potential as therapeutics. Such small molecule therapeutics have significant advantages over non-small molecule agents. Small molecules are more easily delivered into humans or animals than nucleotide analogs such as antagomirs and LNAs. As mentioned previously, small molecules are more stable because they are not susceptible to nucleases *in vivo*, making them more desirable as potential therapeutics. Small molecules are also less likely to induce an immune response than nucleic acid based drugs. Lastly, the pharmacodynamic and pharmacokinetic properties of small molecules are more ideal than nucleic acid based therapeutics [52]. Another major drawback of oligonucleotide-based reagents is the high cost of treatment, while small molecules can generally be manufactured at a lower cost.

Due to the potential of small molecule regulators of miRNAs as both tools to study miRNA biogenesis and as potential therapeutics, several groups have developed assay systems to discover either general or RNA-specific modifiers of miRNA function. The efforts to identify small molecule regulators of miRNAs in cell-based assay systems are discussed in detail. An *in vitro* fluorescence assay that measures the enzymatic activity of Dicer has also been developed that could potentially be used to screen for small molecule modifiers [53].

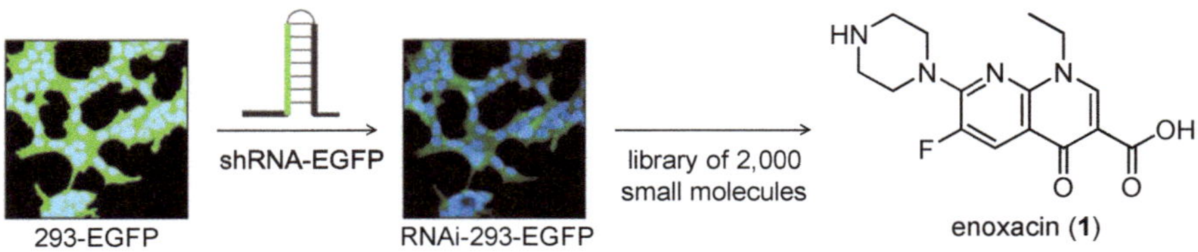

Fig. 2 RNAi-293-EGFP cells show a reduction in fluorescence when compared to 293-EGFP cells. The RNAi-293-EGFP cells were used in an assay for small molecule activators or inhibitors of the small-interfering RNA (siRNA) pathway. In the presence of small molecule activators of the siRNA pathway, a decreased fluorescence is detected, while the presence of a small molecule inhibitor of the siRNA pathway will lead to an increase in fluorescence. The small molecule enoxacin (**1**) was identified as an activator of the RNAi pathway. (Adapted from [54])

3 Small Molecule Regulation of the RNA Interference Pathway

3.1 Small Molecule Activator of the RNA Interference Pathway

Recently, a cell-based assay to monitor the activity of the RNA interference (RNAi) pathway was developed and was used to discover a small molecule, enoxacin (**1**), which enhances siRNA mediated mRNA degradation and promotes the general biogenesis of miRNAs [54]. The assay was developed using a human embryonic kidney (HEK293) cell line that expresses enhanced green fluorescent protein (EFGP). The HEK293 cell line was stably transfected with a lentiviral vector expressing a short hairpin RNA (shRNA) that is processed into a siRNA specifically targeting the EGFP mRNA, generating a cell line termed RNAi-293-EGFP [54]. The expression of the lentivirus encoding the shRNA led to reduced levels of EGFP via the RNAi pathway. From the stably transfected cells, individual cell clones were chosen that had moderate reductions in EGFP expression in order to develop a screen for both inhibitors and activators of the RNAi pathway (Fig. 2). In order to validate the assay system, the RNAi-293-EGFP cells were transfected with a 2′-OMe antisense agent targeting the EGFP siRNA, leading to increased EGFP expression.

The EGFP based assay was utilized in the screening of 2,000 FDA approved compounds and natural products (The Spectrum Collection) in order to identify general activators of the RNAi pathway. The screen identified the small molecule enoxacin (**1,** Fig. 2) that activated siRNA-mediated EGFP degradation leading to reduced fluorescence. Enoxacin reduced EGFP expression approximately threefold at a 50 μM concentration (Fig. 3) and had an EC_{50} of about 30 μM. Cells stably expressing an shRNA targeting an endogenous gene, glyceraldehydes-3-phosphate dehydrogenase (GAPDH), showed similar gene knockdown upon treatment with **1**. The silencing of other mRNAs, including luciferase, through siRNA regulation showed comparable results in the presence of **1**, indicating that the effect of enoxacin

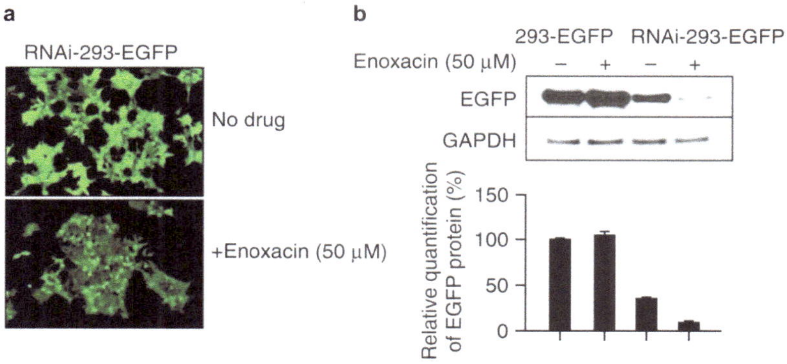

Fig. 3 **a** The activation of siRNA-mediated EGFP degradation in RNAi-293-EGFP cells upon treatment with **1** at 50 μM causing a reduced fluorescence. **b** Western blot analysis of the EGFP protein levels in the RNAi-293-EGFP cells after treatment with **1**. Quantification of the western blot indicates that enoxacin treatment decreased EGFP levels approximately threefold. (Adapted from [54])

on RNAi is universal [54]. Experiments on the gene knockdown efficiency of different concentrations of a transfected siRNA duplex showed that treatment with **1** substantially reduced the siRNA concentration required to achieve comparable knockdown efficiency.

Compounds with a fluoroquinolone core structure [55], similar to enoxacin, were screened for RNAi-enhancing activity to determine if the quinolone family, in general, activates RNAi. Of the nine additional fluoroquinolones assayed, seven showed little to no RNAi-enhancing activity. Two compounds showed substantial activity, however, these two molecules were much less effective than enoxacin. The sensitivity of the RNAi-enhancing activity of enoxacin to chemical modifications indicates that the activity does not depend on the general fluoroquinolone structure, but on the unique chemical structure of enoxacin.

In order to investigate the effect of enoxacin treatment on global gene expression, microarray analyses were performed using HEK293 and NIH3T3 cells. No global effects of **1** on mRNA levels were detected in either cell line, indicating specificity to the components of the RNAi pathway. Further experiments both *in vitro* and *in vivo* revealed more information on the mechanism of action of **1**. In addition to the RNAi-enhancing activity, it was shown that enoxacin can promote the biogenesis of endogenous miRNAs. It was also determined that enoxacin treatment has no effect in an *in vitro* RISC-cleavage assay, indicating that the compound **1** does not affect mRNA-target recognition and cleavage. However, enoxacin was found to promote the processing and loading of siRNAs and miRNAs into RISC. In particular, enoxacin facilitates the interaction between TRBP (TAR RNA binding protein) and RNAs, making the activation of RNAi by enoxacin TRBP dependent. The enhanced interaction between TRBP and RNAs mediated by enoxacin may be the basis of the enhanced siRNA-mediated mRNA degradation observed with enoxacin treatment.

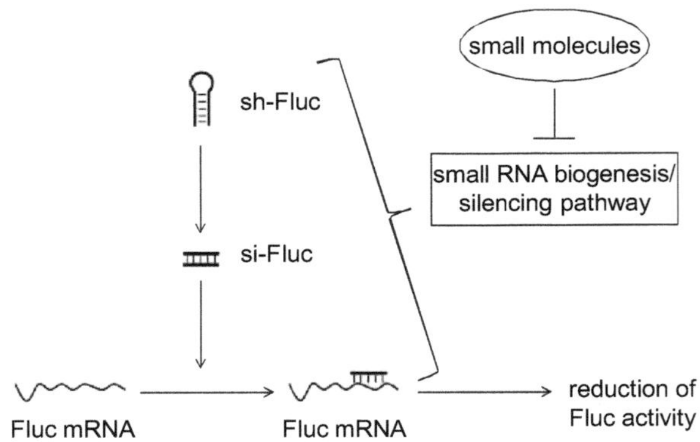

Fig. 4 Assay for small molecule inhibitors of the RNAi pathway via control of firefly luciferase through expression of a short hairpin RNA (shRNA) targeting the firefly luciferase mRNA. In the presence of small molecule inhibitors of the RNAi pathway, an increased luminescence is detected. (Adapted from [56])

3.2 Small Molecule Inhibitors of the RNA Interference Pathway

While enoxacin (**1**) represents an activator of RNAi, molecules that suppress the RNAi pathway were recently identified using a luciferase reporter assay regulated by a simultaneously expressed shRNA (Fig. 4) [56]. The assay employed a firefly luciferase mRNA-targeting shRNA (sh-Fluc) that is processed into a firefly luciferase specific siRNA in the cell. 293T cells were transfected with a firefly luciferase reporter plasmid (pGL3), a *Renilla* luciferase control plasmid (pRL-TK), and the plasmid encoding the firefly luciferase targeting shRNA (pRS-shLuc). Once in the cell, the sh-FLuc is processed into the siRNA which then silences firefly luciferase mRNA produced from the pGL3 plasmid. In order to verify that the decrease in firefly luciferase is induced by siRNA regulation, the same plasmids were transfected into cells which had knocked down Argonaute 2, a key component of RISC and shRNA processing. The firefly luciferase silencing by sh-FLuc was less effective in these cells, indicating that the observed decrease in luminescence is due to RNAi.

The assay was used to screen a library of 530 small molecules in 293T cells transfected with pGL3, pRL-TK, and pRS-shLuc, revealing compound **2** (Fig. 5) that significantly suppressed the siRNA activity and increased the relative luciferase signal by approximately twofold. The inhibition of the shRNA mediated silencing of firefly luciferase by **2** was shown to be dose dependent (Fig. 5). A firefly luciferase Western blot of cells treated with **2** at 3.3 μM for 24 h showed increased protein levels over untreated cells. Because shRNAs and miRNAs follow similar processing pathways in the cell, the compound was assayed for its ability to inhibit miRNA function. Compound **2** was also shown to inhibit miRNA mediated silencing using an assay based on the let-7a regulation of luciferase expression in HeLa cells. A construct containing repeat let-7a target sequences down-stream of a luciferase reporter gene was transfected into HeLa cells along with a let-7a expression plasmid. The overexpression of let-7a reduced luciferase expression, however, upon treatment with **2** the luciferase expression was partially restored, indicating that the small molecule inhibits the let-7a silencing of luciferase.

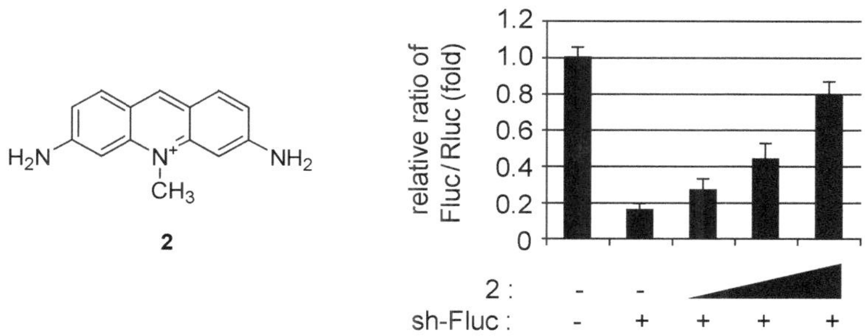

Fig. 5 Trypaflavine (2) is a general inhibitor of the RNAi pathway, as discovered from the screening of 530 small molecules using a firefly luciferase reporter assay based on shRNA regulation. Dose-dependent inhibition of the shRNA silencing of firefly luciferase by **2**. 293T cells were transfected with pGL3, pRL-TK, and pRS-shLuc and were then treated with **2** at 0, 1.5, 3.3, and 7.5 μM. The luciferase activity was quantified after 24 h. In the presence of the small molecule an increased luminescence is detected. (Adapted from [56])

The mechanism by which **2** inhibits siRNA and miRNA mediated gene silencing was further investigated. An *in vitro* Dicer assay measuring Dicer's capacity to cleave double stranded GFP RNA or pre-let-7a showed no change in activity upon treatment with **2**. Instead, inhibition of the siRNA/miRNA pathway by **2** was found to occur through the reduced interaction of siRNAs/miRNAs with Argonaute 2. A precipitation assay for GFP siRNA/Argonaute 2 complexes showed that cells treated with **2** had reduced levels of the siRNA/Argonaute 2 complex, suggesting that **2** inhibits the loading of siRNAs into RISC. Treatment with **2** was also shown to disrupt the association of TRBP and RHA (known cofactors that regulate RNA association with Argonaute 2) with Argonaute 2, which could disrupt the formation of the RISC complex and therefore the loading of siRNAs/miRNAs into RISC [56].

Based on the inhibitory activity of **2** on the RNAi pathway, the compound was assayed for its ability to reverse miRNA-dependent tumorigenesis. The compound was found to reverse the tumorigenicity of miR-93 and miR-130b overexpressing cells [56]. These results support the hypothesis that small molecule inhibitors of miRNAs could lead to potential cancer therapeutics.

Other small molecule inhibitors of the general RNAi pathway have also been identified using a fluorescence reporter assay regulated by an siRNA [57]. Two plasmids expressing RFP and EGFP were transfected into HeLa cells together with an siRNA targeting EGFP mRNA. The assay was used to screen a small chemical library of custom synthesized ATP analogs containing a dihydropteridine scaffold in order to analyze the ATP-dependent steps of the RNAi pathway. The ATP analogs would specifically inhibit ATP-dependent steps in the RNAi pathway. Two compounds were identified, **3** and **4** (Fig. 6), that inhibited the siRNA silencing of EGFP and therefore RNAi. Both compounds were able to completely inhibit siRNA activity at 50 μM (Fig. 6).

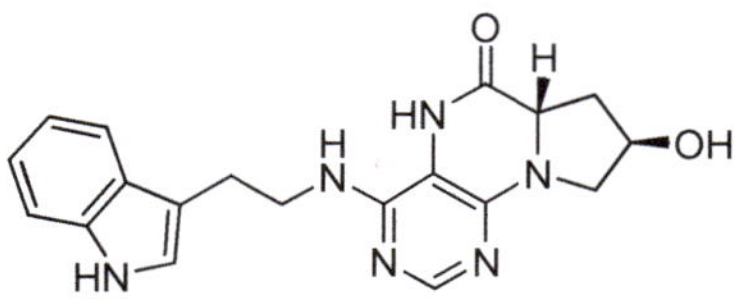

3

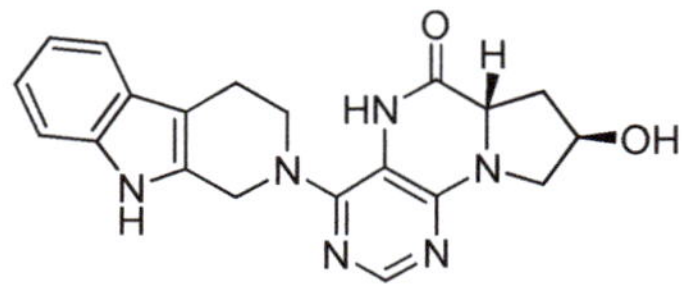

4

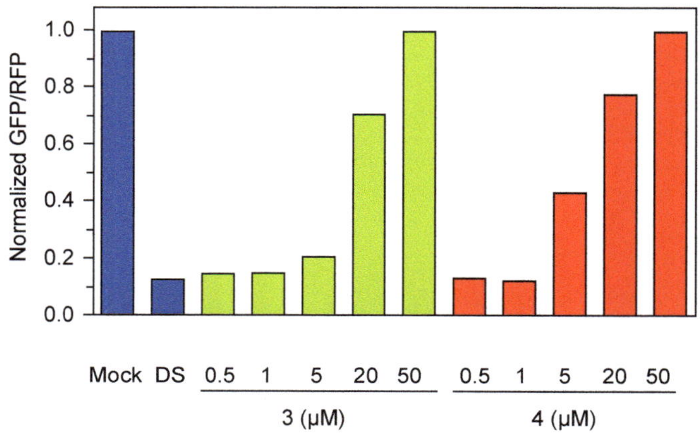

Fig. 6 The general inhibitors **3** and **4** of the RNAi pathway discovered from the screening of ATP analogs using a dual-fluorescence reporter assay based on siRNA regulation. The dual-fluorescence assay results are shown for HeLa cells transfected with the RFP and EGFP plasmids together with 25 nM of the double stranded EGFP siRNA (DS) and treated with **3** or **4** at 0.5–50 μM. Both compounds were able to inhibit siRNA activity in a dose dependent fashion and completely inhibited siRNA activity at 50 μM. (Adapted from [57])

The mode of action of compound **4** was further investigated and it was determined that **4** specifically affects an early unwinding step in the RNAi pathway, but has no effect on RISC formation. It was also discovered that the inhibitor **4** was specific to double stranded siRNA and had no effect on the miRNA let-7 [57].

4 Small Molecule Regulation of miR-16

4.1 MicroRNA miR-16

Thus far, the small molecules discussed have only been identified to be general modifiers of the RNAi and miRNA pathway and are not specific to any one miRNA. More recently, the discovery of small molecule modifiers was directed toward spe-

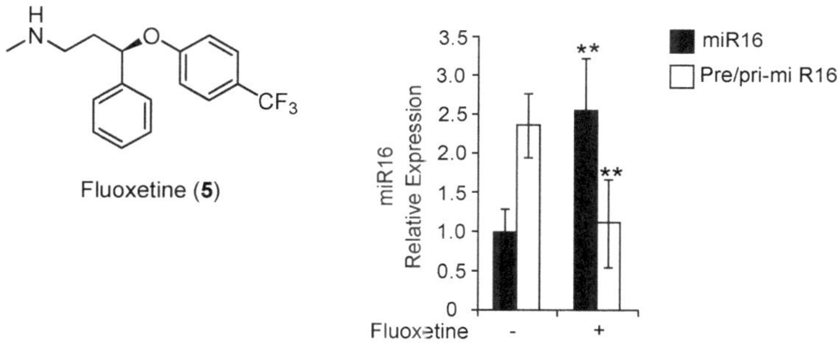

Fig. 7 The SSRI fluoxetine (Prozac, **5**) was shown to increase miR-16 in the serotonergic raphe nuclei region of the mouse brain. Mature miR-16 levels were increased by 2.5-fold in mice raphe after receiving chronic injections (2 μL/min) of **5** (1 μM) for 3 days. Interestingly, the fluoxetine treatment also resulted in decreased levels of pre/pri-miR-16. (Adapted from [61])

cific miRNAs, e.g., miR-16, miR-21, and miR-122. MicroRNA miR-16 is transcribed from the *mir-16-1* gene at chromosome 13q14, a region that is deleted in more than half of B-cell chronic lymphocytic leukemia, one of the most common forms of adult leukemia [58]. Deletions within the 13q14 locus also occur in 50 % of mantle cell lymphoma, 16–40 % of multiple myeloma, and 60 % of prostate cancers [58, 59]. It was reported that miR-16 targets the anti-apoptoic gene *BCL2*, which is overexpressed in many types of human cancers including leukemias and lymphomas [60]. Transfection of miR-16 amd miR-15 into MEG-01 cells, a leukemia derived cell line, led to the down-regulation of *BCL2* expression and the induction of apoptosis [60]. MiRNA miR-15 has also been shown to target serotonin transporter (SERT) mRNAs. A miR-16 target site was predicted in the 3′-UTR of the SERT mRNA and the target was verified by a luciferase reporter assay [61].

4.2 Small Molecule Activator of miR-16

Selective serotonin reuptake inhibitors (SSRIs) are commonly used in the treatment of neuropsychiatric conditions and it has been shown that SSRIs promote reductions in SERT protein levels but have no affect on SERT mRNA levels, which may be attributed to miRNA-mediated inhibition of translation [61]. Because miR-16 was shown to regulate SERT, the affect of small molecule SSRIs including fluoxetine (**5**, Fig. 7) on miR-16 levels was investigated [61].

Experiments in the serotonergic raphe nuclei region of the mouse brain showed a 2.5-fold increase in miR-16 by quantitative RT-PCR upon treatment with chronic injections of fluoxetine (1 μM, 2 μL/min) over 3 days (Fig. 7). The results indicate that fluoxetine may regulate SERT expression through the regulation of miR-16. The fluoxetine treatment in raphe also showed a decrease in pre/pri-miR-16 in addition to the increased levels of mature miR-16, indicating that fluoxetine may

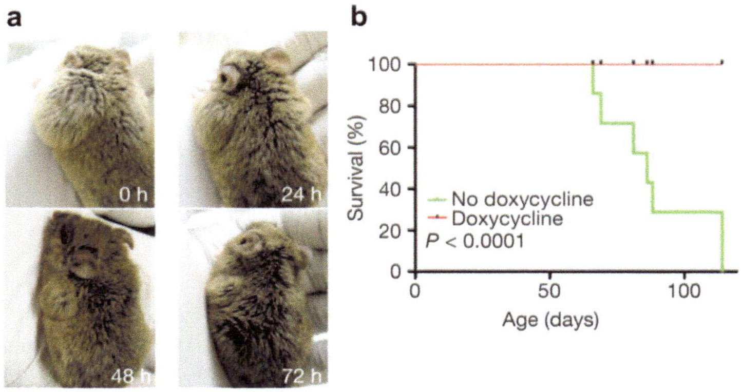

Fig. 8 *NesCre8 mir*-21$^{LSL\text{-}Tetoff}$ mice were engineered to conditionally express miR-21 in the absence of doxycyline. *NesCre8 mir*-21$^{LSL\text{-}Tetoff}$ mice were taken off doxycycline at birth and within 2 months, adults developed signs of lymphoma such as swollen lymph nodes (lymphadenopathy). After the development of signs of lymphoma, the mice were fed doxycycline to turn off miR-21 expression leading to recovery in 100 % of the mice. **a** Time course remission of adenopathy in the lymph nodes of a *NesCre8 mir*-21$^{LSL\text{-}Tetoff}$ mouse before and after treatment with doxycycline. **b** Survival curve of *NesCre8 mir*-21$^{LSL\text{-}Tetoff}$ mice treated with doxycycline after development of lymphoma compared to untreated mice. (Adapted from [16])

activate miR-16 maturation [61]. Based on their results, the authors propose that miR-16 contributes to the therapeutic effect of **5** [61].

5 Small Molecule Regulation of miR-21

5.1 MicroRNA miR-21

MicroRNA miR-21 has been linked to several human cancers. In particular, the over expression of miR-21 has been observed in glioblastomas, breast, pancreatic, cervical, colorectal, ovarian, and lung cancers [62]. miR-21 has been documented to function as an anti-apoptotic factor in cancer cells and it has been demonstrated that knockdown of miR-21 with antisense agents induces apoptosis in glioblastoma [63], breast cancer [64, 65], and hepatocellular carcinoma [66] cell lines. The antitumor activity of miR-21 knockdown was confirmed in mouse xenograft models of glioblastoma with the use of LNA-anti-miR-21 oligonucleotides [67] and of breast cancer upon treatment with miR-21 antagomirs [65], both of which showed a substantial reduction in tumor growth. In mice generated to conditionally express miR-21, *NesCre8 mir*-21$^{LSL\text{-}Tetoff}$, it was shown that overexpression of miR-21 leads to pre-B cell lymphoma and that inactivation of miR-21 leads to regression of the tumors after only a few days as a result of apoptosis (Fig. 8) [16]. These results demonstrate that the malignant phenotype is dependent on the miR-21 levels [16].

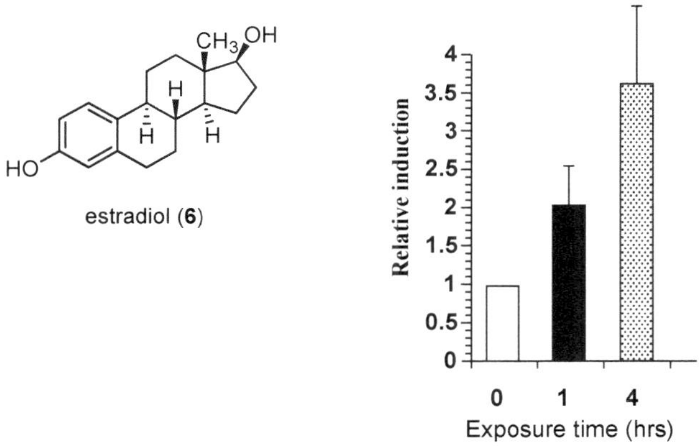

Fig. 9 The natural compound estradiol (**6**) was found to be a modifier of microRNA miR-21 expression in human breast cancer cell lines. Quantitative RT-PCR of miR-21 in MCF-7p cells following exposure to **6** at 10 nM for 0, 1, or 4 h. The small RNA RNU19 was used as a control for normalization. (Adapted from [70])

The involvement of miR-21 in human malignancies and the promising *in vivo* results make miR-21 a potential target for the development of fundamentally new cancer therapeutics.

5.2 *Natural Small Molecule Modifiers of miR-21*

As mentioned previously, the upregulation of miR-21 has been linked to breast cancer and it has been shown that miR-21 levels are significantly higher in estrogen receptor α positive (ERα+) breast cancers than in estrogen receptor α negative (ERα–) breast cancers [68]. Since exposure to estrogens has been widely accepted as a major risk factor for the development of breast cancer, investigations on the effect of estradiol (**6**, Fig. 9) on miR-21 expression have been conducted. Estradiol functions as a ligand for ERα and ERβ, inducing a transduction of the ligand-receptor complex into the nucleus followed by the activation of gene transcription. Interestingly, two studies came to opposite conclusions on the effect of estradiol on miR-21 levels in the ERα+ human breast cancer cells line MCF-7, either linking miR-21 downregulation [69] or miR-21 upregulation [70] to the exposure of cells to **6**. A recent study reports a downregulation of miR-21 following exposure of MCF-7 cells to estradiol at a 10 nM concentration for 6 h [69]. Quantitative RT-PCR (qRT-PCR) experiments showed a reduction of 60 % in mature miR-21 levels caused by exposure to **6**. In order to further explore the downregulation of miR-21 by **6**, experiments on the expression of the miR-21 target genes were conducted by measuring the mRNA and protein levels of the endogenous targets PDCD4, PTEN, and BCL2 using qRT-PCR and Western blot analysis. As expected, exposure to **6** led to increased mRNA and protein levels of PDCD4 and BCL2, and increased protein expression of PTEN. Exposure of cells to a combination of estradiol and 4-hydroxytamoxifen, the active metabolite of the anti-estrogen tamoxifen, or fulvestrant (ICI 182780), a known antagonist of ERα, led to increased miR-21 levels compared to cells exposed to only **6**, indicating that the effect of estradiol on miR-21 levels is ERα dependent [69].

In contrast, a second study on the effect of estradiol on miR-21 levels in breast cancer cells came to the conclusion that miR-21 expression is upregulated in the presence **6**. Estradiol increased miR-21 levels in MCF-7 cells and cell lines derived from MCF-7 by twofold after a 4 h exposure at 10 nM (Fig. 9) [70]. It was further demonstrated that treatment with **6** induced the expression of 20 other miRNAs, including the let-7 family, and inhibited seven other miRNAs. To determine the requirement of ERα for estradiol induced miR-21 expression, MCF-7 cells were pre-treated with fulvestrant (ICI 182780) for 24 h which causes ERα degradation, followed by exposure to **6** for 4 h. The levels of miR-21 were elevated in cells pretreated with fulvestrant, but were only marginally increased with treatment of **6**, suggesting that free ERα represses miR-21 expression and that estradiol increases miR-21 expression by relieving the repression caused by free ERα. It was further shown that ERα binding sites were located in the miR-21 regulatory region and that Dicer mRNA was induced by estradiol treatment, suggesting that estradiol regulates the miRNA pathway on both the transcriptional level as well as miRNA processing [70]. These two studies offer very different conclusions on the estradiol regulated expression of miR-21 in breast cancer. Although the reasons behind the discrepancy are unknown, the opposing results reveal the complexity of miRNA biogenesis and regulation.

The natural product nicotine, which promotes gastric tumor growth, has recently been shown to upregulate miR-21 and miR-16 in gastric cancer cells [71]. The expression profiles of 95 cancer related miRNAs were examined in untreated and nicotine-treated human gastric adenocarcinoma cells, revealing 3.2 and 2.7-fold increases in the expression of miR-21 and miR-16, respectively. The increased miRNA expression seems to be a result of enhanced binding of nuclear factor kappa B (NF-κB) to the promoters of miR-21 and miR-16 through nicotine exposure [71].

5.3 Activation of miR-21 by the Anticancer Agent 5-Fluorouracil

Given that miRNAs are involved in the initiation and progression of cancers, it was hypothesized that anti-cancer drugs could affect miRNA expression levels since several of these compounds are able to interfere with nucleic acid metabolism and gene expression. The anti-cancer drug 5-fluorouracil (**7**, Fig. 10), which has been widely used in the treatment of colon, breast, and pancreatic cancers [72], was tested for its affect on the expression of 153 miRNAs in colon cancer cells [73].

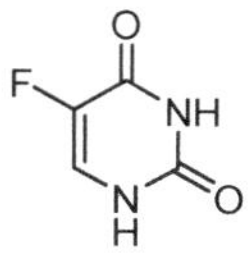

5-fluorouracil (7)

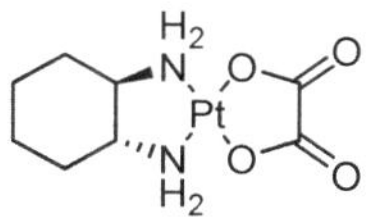

oxaliplatin (8)

Fig. 10 The structures of the chemotherapeutics 5-fluorouracil (**7**) and oxaliplatin (**8**), which were found to be modifiers of several miRNAs, including miR-21 in human colon cancer cell lines

The exposure of two colon cancer cell lines, C22.20 and HC.21, to **7** at 10 μM for 6 days led to a twofold to eightfold increase in the expression of 17 miRNAs, including miR-21. In addition, three miRNAs (miR-200b, 210, and 224) were down-regulated in the presence of **7** [73]. The up-regulation of anti-apoptotic miRNAs, like miR-21 and miR-19a (a member of the miR-17-92 cluster), identified in this study led the authors hypothesize that the miRNA-response represents a cellular defense mechanism against the drug treatment. Apoptosis induced by **7** was examined in the colorectal cancer cell lines Colo-320DM and SW620 and it was shown that transfection of miR-21 led to reduced levels of apoptosis and cell-cycle arrest [74]. Xenograft experiments in nude mice overexpressing miR-21 also resulted in a reduced response to treatment with **7** [74]. A miR-21 resistance to the anti-tumor effect of a combination therapy of **7** and interferon-α has also been demonstrated in hepatocellular carcinoma cells [75].

Another study in the colon cancer cell lines HCT-8 and HCT-116 also showed altered expression profiles of miRNAs upon treatment with **7** and upon treatment with the chemotherapeutic oxaliplatin (**8**, Fig. 10) [76]. It has further been shown that 5-fluorouracil (**7**) induces significantly altered expression profiles of 42 miRNAs in MCF-7 breast cancer cells after 48 h at 10 nM [77]. Of these miRNAs, 23 were up-regulated and 19 were down-regulated and it was shown that several of the cellular targets of these misregulated miRNAs are either oncogenes or tumor suppressors [77].

5.4 Discovery of Inhibitors of miR-21 from Small Molecule Screens

Due to the biological relevance of miR-21 as an anti-apoptotic factor in cancer cells and its elevated levels in various cancers, much interest has been placed on its regulation. Recently, the first synthetic small molecule inhibitors of miR-21 were discovered using a lentiviral reporter assay and screening approach [78]. A lentiviral reporter construct for miR-21 was assembled by introducing the complementary sequence to the mature miR-21 downstream of a luciferase reporter gene as shown in Fig. 11. The construct serves as a sensor for detecting endogenous mature miR-21. The luciferase reporter, Luc-miR-21, was stably transfected into the human cervical

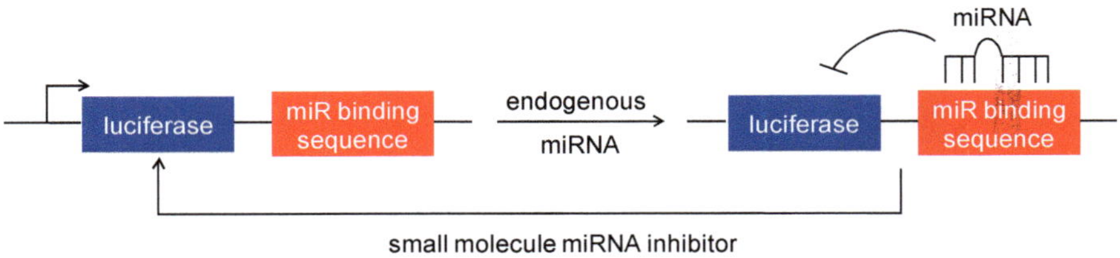

Fig. 11 Assay for miRNA function via control of luciferase expression by a miRNA target sequence in the 3′ untranslated region. The presence of the complementary miRNA in cells transfected with the reporter construct leads to decreased luciferase expression. In the presence of a small molecule inhibitor of miRNA function, the luciferase is expressed and an increase in luminescence is detected

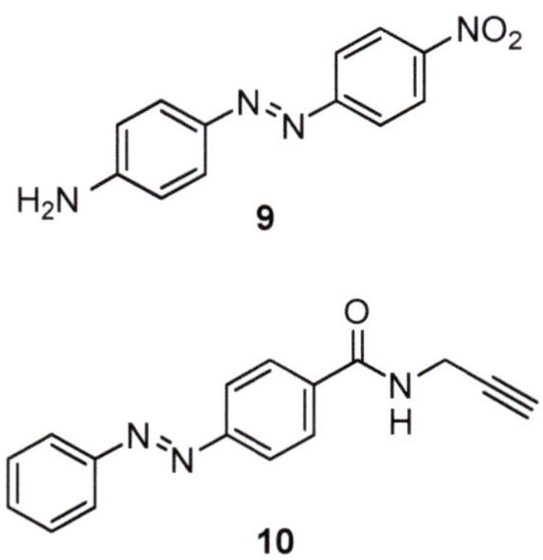

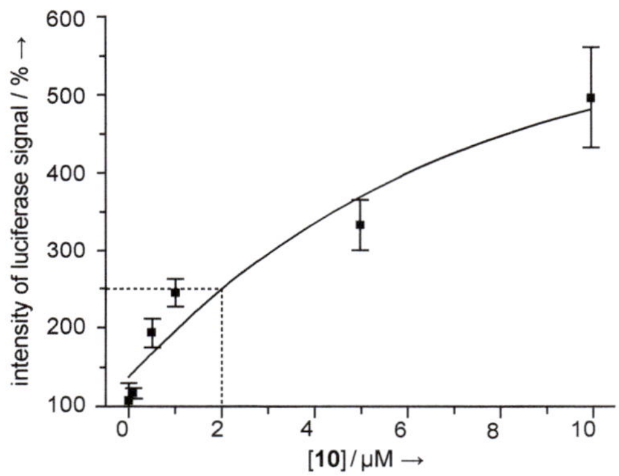

Fig. 12 Inhibitors **9** and **10** of microRNA miR-21 discovered from the screening of a library of small molecules using a luciferase based assay system in HeLa cells. Dose response curve for **10** in HeLa cells stably transfected with a Luc-miR-21 reporter. The results show a fivefold increase in luciferase signal after treatment with **10** at 10 µM and an EC_{50} of 2 µM. (Adapted from [78])

cancer cell line HeLa. The high level of endogenous mature miR-21 in HeLa cells [79] led to a 90 % decreased luciferase signal in comparison to a control Luc-linker construct that contained no known miRNA target sequence.

The HeLa cells stably transfected with the Luc-miR-21 reporter were used to screen an initial set of more than 1,000 small molecules, including the commercially available LOPAC collection, at 10 µM. An initial hit compound, the diazobenzene **9**, was discovered, that led to a 2.5-fold increase in the luciferase signal over control cells only exposed to DMSO. A preliminary structure-activity-relationship study and further screening of approximately 100 structurally similar molecules to the diazobenzene core structure of **9** delivered the more active compound **10** (Fig. 12). At a concentration of 10 µM, **10** induces a fivefold increase in luciferase signal and has an EC_{50} of 2 µM (Fig. 12). HeLa cells stably expressing a control reporter with no miRNA target sequence downstream of luciferase showed no effect on the luminescence signal when treated with **10**, indicating that the increased luciferase in the Luc-miR-21 assay is caused by inhibition of the miRNA pathway and not general luciferase inhibition.

Quantitative RT-PCR experiments to measure the intracellular miRNA levels showed that treatment with **10** led to a reduction of 78 % in miR-21 expression in HeLa cells relative to DMSO treatment. Similar experiments measuring the expression of endogenous miR-93 and exogenous miR-30 showed no reduction of miRNA levels upon treatment with **10**, revealing that **10** is not a general inhibitor of the miRNA pathway but has some specificity for miR-21. Similar results were obtained after treatment with **10** in other cell lines which are known to have high levels of miR-21 including MCF-7 (breast cancer), MDA-MB-231 (breast cancer), and A172 (glioblastoma) cells.

In order to further investigate the mechanism of **10**, the intracellular levels of pri-miR-21 after treatment with **10** were quantified and revealed a reduction of 87 % in HeLa cells. This indicates that the diazobenzene **10** inhibits miR-21 at the transcriptional level, but not at any of the downstream processes within the miRNA pathway [78]. The efficient and specific inhibition of miR-21 by **10** provides evidence that small molecule modifiers of miRNA function can be used as probes to investigate the regulation of specific miRNAs and offers the potential for the development of small molecule-based miRNA-targeting therapeutics.

6 Small Molecule Regulation of miR-122

6.1 MicroRNA miR-122

Transcribed from the gene *hcr*, miR-122 is a liver specific miRNA and is the most abundant miRNA in the liver [80]. The most common function of miR-122 in the liver is the regulation of lipid and cholesterol metabolism [81]. Recently, it was discovered that miR-122 is down-regulated in hepatocellular carcinoma (HCC), a primary cancer of the liver. HCC is the third largest cause of cancer related death behind only lung and colon cancers and prognosis are usually poor [82]. The levels of miR-122 in the HCC cell line Huh7 are reduced by ~85 % and by >99 % in the HCC cell lines HepG2 and Hep3B in comparison to healthy liver tissue [80, 83]. Several cellular targets of miR-122 in primary liver carcinomas and the HCC cell lines Hep3B and HegG2 have been identified, including cyclin-G1 (CCNG1) and Bcl-w, an anti-apoptotic Bcl-2 family member [84, 85]. Because of the low levels of miR-122 in HCC, the cellular mRNA and protein levels of Bcl-w are highly expressed, leading to an enhanced viability of cancer cells [83]. In fact, transfection of miR-122 into cancer cells led to reduced levels of Bcl-w and the induction of apoptosis [83].

Moreover, many cases of HCC result from chronic infections with the hepatitis C virus (HCV) [86] and, interestingly, the presence of miR-122 has also been implicated in HCV replication. It was discovered that miR-122 is necessary for HCV replication and infectious virus production through interaction with the viral genome. The HCV genome contains two miR-122 target sites in the 5′ non-coding region of the virus and miR-122 binding results in the up-regulation of viral RNA [87, 88]. Knockdown of miR-122 using a 2′-OMe miR-122 antagomir resulted in a decrease of HCV RNA in human liver cells [87]. Similar studies using LNA oligonucleotides complementary to miR-122 to reduce cellular levels of miR-122 showed decreased HCV levels in chronically infected primates [89]. These promising results using nucleic acid based inhibitors as miR-122 targeting therapeutics provided the rational for a search for miR-122 small molecule inhibitors.

6.2 Discovery of Inhibitors of miR-122 from Small Molecule Screens

Small molecule modifiers of miR-122 expression could be used a molecular probes providing valuable information on the regulation of miR-122 in HCC and could aid in the development of novel therapeutics for HCC and HCV. A reporter construct was developed to identify small molecule modifiers of miR-122 function [90]. The construct expresses both *Renilla* and firefly luciferase allowing for an internal control for cell number and viability. The complementary sequence for mature miR-122 was introduced downstream of the *Renilla* luciferase gene to create the reporter construct psiCHECK-miR122. Endogenous miR-122 can bind the target sequence and lead to a decrease in the *Renilla* luciferase signal in a liver cell line (Fig. 11). When transfected into Huh7 cells, a human hepatoma cell line, the luciferase signal was reduced about tenfold when compared to a control construct without the miR-122 target sequence. In contrast, when transfected into HeLa cells where miR-122 is not expressed, the luciferase signal was the same for both psiCHECK-miR122 and the control construct, verifying that the psiCHECK-miR122 reporter is a cellular sensor for endogenous miR-122. In order to validate that this reporter construct can be employed in assays for small molecule modifiers of miR-122 function, Huh7 cells were co-transfected with psiCHECK-miR122 and a 2′-OMe miR-122 antagomir. The presence of the miR-122 antagomir restored the luciferase signal, indicating that the reporter construct could be used in a cellular assay to identify small molecule modifiers of miR-122.

Over 1,300 compounds from the NCI Diversity Set II were screened for an increase of the *Renilla* luciferase signal and two inhibitors of miR-122, the sulfonamide **11** and the amide **12** (Fig. 13), were identified. At a 10 μM concentration in Huh7 cells, **11** induced a 12-fold increase and **12** induced an approximately eightfold increase in the relative luciferase signal over cells only exposed to a DMSO control.

A structure-activity relationship study of each of the miR-122 inhibitors was performed, showing that chemical modifications of the original structures often led to decreased activity or even complete loss of activity. In a dose-dependent

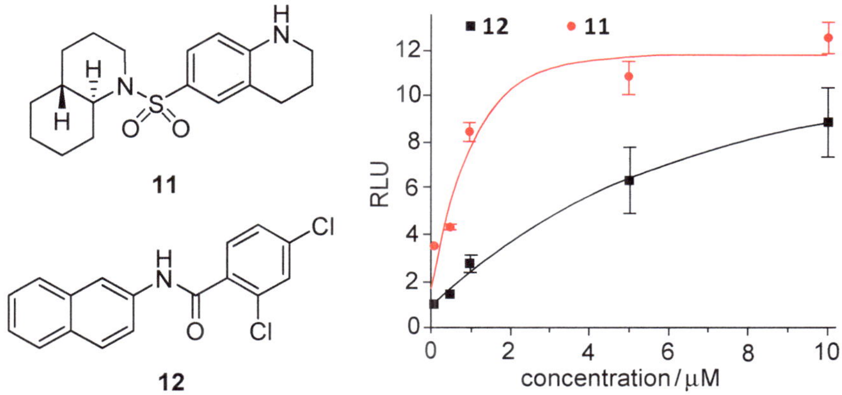

Fig. 13 Inhibitors **11** and **12** of microRNA miR-122 discovered from the screening of a library of small molecules using a luciferase based assay system in Huh7 cells. Dose response curves for **11** and **12** in Huh7 cells transfected with the psiCHECK-miR122 reporter. EC_{50} values of 0.6 and 3 μM were determined for **11** and **12**, respectively. (Adapted from [90])

assay, **11** and **12** were found to have EC_{50} values of 0.6 and 3 μM, respectively (Fig. 13). Quantitative RT-PCR experiments measuring the intracellular miRNA levels showed that treatment of Huh7 cells with **11** (10 μM) led to a reduction of 72 % in miR-122 expression relative to DMSO treatment. Exposure of Huh7 cells to **12** (10 μM) induced a reduction in miR-122 levels of 45 % relative to DMSO. Similar experiments measuring the levels of miR-21 in HeLa cells showed no reduced expression upon treatment with **11** and **12**, revealing that the compounds are not general inhibitors of the miRNA pathway but rather display specificity for miR-122. Quantitative RT-PCR measuring the intracellular levels of pri-miR-122 in Huh7 cells after treatment with **11** or **12** also revealed reductions of 97 and 78 %, respectively, indicating that both inhibitors are most likely targeting the transcription of the miR-122 gene into pri-miR-122, although the exact mechanism and target of the compounds is unknown [90].

As mentioned previously, the inhibition of miR-122 by antisense oligonucleotides results in the reduction of HCV replication in Huh7 cells [87]. Because of the efficient downregulation of miR-122 by **11** and **12**, the small molecules were assayed for their effect on HCV replication in Huh7 cells. Huh7 cells transfected with genotype 1a H77c RNA [91] were either transfected with a 2′-OMe miR-122 antagomir (positive control) or were treated with **11** or **12**. In agreement with previous reports, the 2′-OMe antagomir reduced HCV RNA levels by 80 % [87, 88]. Treatment with the small molecule inhibitors caused reductions in HCV RNA levels of 53 and 52 % for **11** and **12**, respectively. These results demonstrate that small molecule inhibitors of miR-122 have potential as new therapeutics for the treatment of HCV infection.

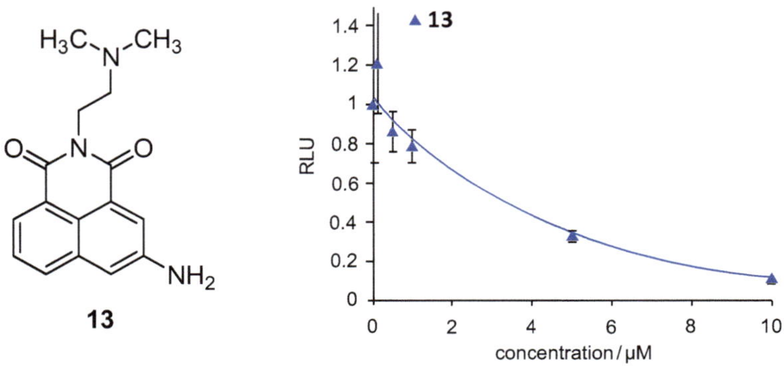

13

Fig. 14 Activator **13** of microRNA miR-122 discovered from the screening of a library of small molecules using a luciferase based assay system in Huh7 cells. Dose response curve for the miR-122 activator **13** in Huh7 cells transfected with the psiCHECK-miR122 reporter. An IC_{50} value of 3 μM was determined for **13**. (Adapted from [90])

6.3 Discovery of an Activator of miR-122 from Small Molecule Screens

As discussed previously, compared to healthy liver tissue, miR-122 is reduced by 85 % in the HCC cell line Huh7 and by 99 % in the HCC cell lines HepG2 and Hep3B [80, 83]. Thus, the screening results of the NCI Diversity Set II compounds in Huh7 cells transfected with psiCHECK-miR122 were also analyzed for a further reduction in the relative *Renilla* luciferase signal, since this could reveal small molecules that would activate miR-122 function in Huh7 cells [90]. Compound **13** (Fig. 14) was found to be an activator of miR-122, inducing a sevenfold reduction in the *Renilla* luciferase signal.

In a dose-dependent assay, **13** was found to have an IC_{50} value of 3 μM (Fig. 14). Quantitative RT-PCR experiments measuring the intracellular miR-122 levels showed that treatment of Huh7 cells with **13** (10 μM) led to a 438 % increase in miR-122 expression relative to DMSO treatment. Similar experiments measuring the levels of miR-21 in HeLa cells showed no increased expression upon treatment with **13**, revealing that the compound is not a general activator of the miRNA pathway but displays some level of specificity for miR-122. Quantitative RT-PCR measuring the intracellular levels of pri-miR-122 in Huh7 cells after treatment with **13** also showed up-regulation of pri-miR122, suggesting that the small molecule targets miR-122 transcription, although the exact mechanism and target of the compound is unknown [90].

As mentioned previously, miR-122 is greatly down-regulated in HCC compared to healthy liver tissue which results in the upregulation of Bcl-w, an antiapoptotic target of miR-122. The high expression levels of Bcl-w leads to a deactivation of caspase-3 and an enhanced viability of cancer cells [83]. Because of the efficient activation of miR-122 by **13**, the small molecule was assayed for its ability to induce apoptosis in the HCC cell line HepG2. Treatment of HepG2 cells with **13** led to an approximately 20-fold increase in the activity of caspase-3 and -7 over cells treated with a DMSO control [90]. The activator **13** was also successful at reducing cell viability by approximately 80 % in HepG2 cells, but had little effect on Huh7 cells, indicating that there is a selective effect in cells with pathologically low levels of miR-122 [90].

7 Conclusion

Regulation of miRNAs in both a general and a miRNA-specific fashion has been observed in response to the treatment of cells and model organisms with naturally occurring or synthetic small molecules. Through several cellular based assay systems, small molecule regulators of the miRNA pathway have been identified. These small molecule modifiers represent unique tools to further study the regulation and biogenesis of miRNAs. In addition, because of the implication of miRNAs in various human malignancies, small molecules which regulate miRNA expression could provide insight into the misregulation of miRNAs in diseases. These small molecules also validate the miRNA pathway as a potential target for existing therapeutics and the development of new therapeutics.

References

1. Lee R, Feinbaum R, Ambros V (1993) The C. elegans heterochronic gene lin-4 encodes small RNAs with antisense complementarity to lin-14. Cell 75:843–854
2. Reinhart B, Slack F, Basson M, Pasquinelli A, Bettinger J et al (2000) The 21-nucleotide let-7 RNA regulates developmental timing in Caenorhabditis elegans. Nature 403:901–906
3. Pasquinelli A, Reinhart B, Slack F, Martindale M, Kuroda M et al (2000) Conservation of the sequence and temporal expression of let-7 heterochronic regulatory RNA. Nature 408:86–89
4. Winter J, Jung S, Keller S, Gregory R, Diederichs S (2009) Many roads to maturity: microRNA biogenesis pathways and their regulation. Nat Cell Biol 11:228–234
5. Ghildiyal M, Zamore P (2009) Small silencing RNAs: an expanding universe. Nat Rev Genet 10:94–108
6. Carthew R, Sontheimer E (2009) Origins and Mechanisms of miRNAs and siRNAs. Cell 136:642–655
7. Garzon R, Marcucci G, Croce C (2010) Targeting microRNAs in cancer: rationale, strategies and challenges. Nat Rev Drug Discov 9:775–789
8. Shenouda S, Alahari S (2009) MicroRNA function in cancer: oncogene or a tumor suppressor? Cancer Metastasis Rev 28:369–378
9. Esquela-Kerscher A, Slack F (2006) Oncomirs—microRNAs with a role in cancer. Nat Rev Cancer 6:259–269
10. Carthew R (2006) Gene regulation by microRNAs. Curr Opin Genet Dev 16:203–208
11. Vasudevan S, Tong Y, Steitz J (2007) Switching from repression to activation: microRNAs can up-regulate translation. Science 318:1931–1934

12. Vasudevan S, Tong Y, Steitz JA (2008) Cell-cycle control of microRNA-mediated translation regulation. Cell Cycle 7:1545–1549
13. Appasani K (2008) MicroRNAs: from basic science to disease biology. Cambridge University Press, Cambridge
14. Sevignani C, Calin G, Siracusa L, Croce C (2006) Mammalian microRNAs: a small world for fine-tuning gene expression. Mamm Genome 17:189–202
15. Zhang S, Chen L, Jung E, Calin G (2010) Targeting MicroRNAs With Small Molecules: From Dream to Reality. Clin Pharmacol Ther 87:754–758
16. Medina PP, Nolde M, Slack FJ (2010) OncomiR addiction in an *in vivo* model of microRNA-21-induced pre-B-cell lymphoma. Nature 467:86–90
17. Du T, Zamore P (2005) microPrimer: the biogenesis and function of microRNA. Development 132:4645–4652
18. Lee Y, Ahn C, Han J, Choi H, Kim J et al (2003) The nuclear RNase III Drosha initiates microRNA processing. Nature 425:415–419
19. Bernstein E, Caudy A, Hammond S, Hannon G (2001) Role for a bidentate ribonuclease in the initiation step of RNA interference. Nature 409:363–366
20. Grishok A, Pasquinelli A, Conte D, Li N, Parrish S et al (2001) Genes and mechanisms related to RNA interference regulate expression of the small temporal RNAs that control C. elegans developmental timing. Cell 106:23–34
21. Hutvágner G, McLachlan J, Pasquinelli A, Bálint E, Tuschl T et al (2001) A cellular function for the RNA-interference enzyme Dicer in the maturation of the let-7 small temporal RNA. Science 293:834–838
22. Ketting R, Fischer S, Bernstein E, Sijen T, Hannon G et al (2001) Dicer functions in RNA interference and in synthesis of small RNA involved in developmental timing in C. elegans. Genes Dev 15:2654–2659
23. Yi R, Qin Y, Macara I, Cullen B (2003) Exportin-5 mediates the nuclear export of pre-microRNAs and short hairpin RNAs. Genes Dev 17:3011–3016
24. Pillai RS (2005) MicroRNA function: multiple mechanisms for a tiny RNA? RNA 11:1753–1761
25. He L, Hannon G (2004) MicroRNAs: small RNAs with a big role in gene regulation. Nat Rev Genet 5:522–531
26. Bartel D (2004) MicroRNAs: genomics, biogenesis, mechanism, and function. Cell 116:281–297
27. Chan S, Slack F (2006) microRNA-mediated silencing inside P-bodies. RNA Biol 3:97–100
28. Lelandais-Brière C, Sorin C, Declerck M, Benslimane A, Crespi M et al (2010) Small RNA diversity in plants and its impact in development. Curr Genomics 11:14–23
29. Krol J, Loedige I, Filipowicz W (2010) The widespread regulation of microRNA biogenesis, function and decay. Nat Rev Genet 11:597–610
30. Zhang X, Zeng Y (2010) Regulation of mammalian microRNA expression. J Cardiovasc Transl Res 3:197–203
31. He L, Thomson J, Hemann M, Hernando-Monge E, Mu D et al (2005) A microRNA polycistron as a potential human oncogene. Nature 435:828–833
32. Calin G, Sevignani C, Dumitru C, Hyslop T, Noch E et al (2004) Human microRNA genes are frequently located at fragile sites and genomic regions involved in cancers. Proc Natl Acad Sci U S A 101:2999–3004
33. Calin GA, Dumitru CD, Shimizu M, Bichi R, Zupo S et al (2002) Frequent deletions and down-regulation of micro- RNA genes miR15 and miR16 at 13q14 in chronic lymphocytic leukemia. Proc Natl Acad Sci U S A 99:15524–15529
34. Saito Y, Liang G, Egger G, Friedman J, Chuang J et al (2006) Specific activation of microRNA-127 with downregulation of the proto-oncogene BCL6 by chromatin-modifying drugs in human cancer cells. Cancer Cell 9:435–443
35. Zhang L, Huang J, Yang N, Greshock J, Megraw M et al (2006) microRNAs exhibit high frequency genomic alterations in human cancer. Proc Natl Acad Sci U S A 103:9136–9141

36. Shi X, Tepper C, deVere White R (2008) Cancerous miRNAs and their regulation. Cell Cycle 7:1529–1538
37. O'Donnell K, Wentzel E, Zeller K, Dang C, Mendell J (2005) c-Myc-regulated microRNAs modulate E2F1 expression. Nature 435:839–843
38. Kim J, Inoue K, Ishii J, Vanti W, Voronov S et al (2007) A MicroRNA feedback circuit in midbrain dopamine neurons. Science 317:1220–1224
39. Thomson J, Newman M, Parker J, Morin-Kensicki E, Wright T et al (2006) Extensive post-transcriptional regulation of microRNAs and its implications for cancer. Genes Dev 20:2202–2207
40. Davis B, Hilyard A, Lagna G, Hata A (2008) SMAD proteins control DROSHA-mediated microRNA maturation. Nature 454:56–61
41. Trabucchi M, Briata P, Garcia-Mayoral M, Haase A, Filipowicz W et al (2009) The RNA-binding protein KSRP promotes the biogenesis of a subset of microRNAs. Nature 459:1010–1014
42. Viswanathan S, Daley G (2010) Lin28: A microRNA regulator with a macro role. Cell 140:445–449
43. Hagan J, Piskounova E, Gregory R (2009) Lin28 recruits the TUTase Zcchc11 to inhibit let-7 maturation in mouse embryonic stem cells. Nat Struct Mol Biol 16:1021–1025
44. Heo I, Joo C, Cho J, Ha M, Han J et al (2008) Lin28 mediates the terminal uridylation of let-7 precursor MicroRNA. Mol Cell 32:276–284
45. Esau C (2008) Inhibition of microRNA with antisense oligonucleotides. Methods 44:55–60
46. Veedu R, Wengel J (2010) Locked nucleic acids: promising nucleic acid analogs for therapeutic applications. Chem Biodivers 7:536–542
47. Brown B, Naldini L (2009) Exploiting and antagonizing microRNA regulation for therapeutic and experimental applications. Nat Rev Genet 10:578–585
48. Liu Z, Sall A, Yang D (2008) MicroRNA: An emerging therapeutic target and intervention tool. Int J Mol Sci 9:978–999
49. Grünweller A, Hartmann R (2007) Locked nucleic acid oligonucleotides: the next generation of antisense agents? BioDrugs 21:235–243
50. Ebert M, Neilson J, Sharp P (2007) MicroRNA sponges: competitive inhibitors of small RNAs in mammalian cells. Nat Methods 4:721–726
51. Carè A, Catalucci D, Felicetti F, Bonci D, Addario A et al (2007) MicroRNA-133 controls cardiac hypertrophy. Nat Med 13:613–618
52. Zhang S, Chen L, Jung E, Calin G (2010) Targeting microRNAs with small molecules: from dream to reality. Clin Pharmacol Ther 87:754–758
53. Davies B, Arenz C (2006) A homogenous assay for micro RNA maturation. Angew Chem Int Ed Engl 45:5550–5552
54. Shan G, Li Y, Zhang J, Li W, Szulwach K et al (2008) A small molecule enhances RNA interference and promotes microRNA processing. Nat Biotechnol 26:933–940
55. Bhanot SK, Singh M, Chatterjee NR (2001) The chemical and biological aspects of fluoroquinolones: reality and dreams. Curr Pharm Des 7:311–335
56. Watashi K, Yeung M, Starost M, Hosmane R, Jeang K (2010) Identification of small molecules that suppress microRNA function and reverse tumorigenesis. J Biol Chem 285:24707–24716
57. Chiu Y, Dinesh C, Chu C, Ali A, Brown K et al (2005) Dissecting RNA-interference pathway with small molecules. Chem Biol 12:643–648
58. Calin G, Dumitru C, Shimizu M, Bichi R, Zupo S et al (2002) Frequent deletions and down-regulation of micro- RNA genes miR15 and miR16 at 13q14 in chronic lymphocytic leukemia. Proc Natl Acad Sci U S A 99:15524–15529
59. Dong JT, Boyd JC, Frierson HF (2001) Loss of heterozygosity at 13q14 and 13q21 in high grade, high stage prostate cancer. Prostate 49:166–171
60. Cimmino A, Calin GA, Fabbri M, Iorio MV, Ferracin M et al (2005) miR-15 and miR-16 induce apoptosis by targeting BCL2. Proc Natl Acad Sci U S A 102:13944–13949

61. Baudry A, Mouillet-Richard S, Schneider B, Launay J, Kellermann O (2010) miR-16 targets the serotonin transporter: a new facet for adaptive responses to antidepressants. Science 329:1537–1541
62. Tong A, Nemunaitis J (2008) Modulation of miRNA activity in human cancer: a new paradigm for cancer gene therapy? Cancer Gene Ther 15:341–355
63. Ciafrè S, Galardi S, Mangiola A, Ferracin M, Liu C et al (2005) Extensive modulation of a set of microRNAs in primary glioblastoma. Biochem Biophys Res Commun 334:1351–1358
64. Iorio M, Ferracin M, Liu C, Veronese A, Spizzo R et al (2005) MicroRNA gene expression deregulation in human breast cancer. Cancer Res 65:7065–7070
65. Si M, Zhu S, Wu H, Lu Z, Wu F et al (2007) miR-21-mediated tumor growth. Oncogene 26:2799–2803
66. Meng F, Henson R, Wehbe-Janek H, Ghoshal K, Jacob S et al (2007) MicroRNA-21 regulates expression of the PTEN tumor suppressor gene in human hepatocellular cancer. Gastroenterology 133:647–658
67. Corsten M, Miranda R, Kasmieh R, Krichevsky A, Weissleder R et al (2007) MicroRNA-21 knockdown disrupts glioma growth *in vivo* and displays synergistic cytotoxicity with neural precursor cell delivered S-TRAIL in human gliomas. Cancer Res 67:8994–9000
68. Mattie M, Benz C, Bowers J, Sensinger K, Wong L et al (2006) Optimized high-throughput microRNA expression profiling provides novel biomarker assessment of clinical prostate and breast cancer biopsies. Mol Cancer 5:24
69. Wickramasinghe N, Manavalan T, Dougherty S, Riggs K, Li Y et al (2009) Estradiol downregulates miR-21 expression and increases miR-21 target gene expression in MCF-7 breast cancer cells. Nucleic Acids Res 37:2584–2595
70. Bhat-Nakshatri P, Wang G, Collins N, Thomson M, Geistlinger T et al (2009) Estradiol-regulated microRNAs control estradiol response in breast cancer cells. Nucleic Acids Res 37:4850–4861
71. Shin VY, Jin H, Ng EK, Cheng AS, Chong WW et al (2010) NF-kappaB targets miR-16 and miR-21 in gastric cancer: involvement of prostaglandin E receptors. Carcinogenesis
72. Longley DB, Harkin DP, Johnston PG (2003) 5-fluorouracil: mechanisms of action and clinical strategies. Nat Rev Cancer 3:330–338
73. Rossi L, Bonmassar E, Faraoni I (2007) Modification of miR gene expression pattern in human colon cancer cells following exposure to 5-fluorouracil *in vitro*. Pharmacol Res 56:248–253
74. Valeri N, Gasparini P, Braconi C, Paone A, Lovat F et al (2010) MicroRNA-21 induces resistance to 5-fluorouracil by down-regulating human DNA MutS homolog 2 (hMSH2). Proc Natl Acad Sci U S A 107:21098–21103
75. Tomimaru Y, Eguchi H, Nagano H, Wada H, Tomokuni A et al (2010) MicroRNA-21 induces resistance to the anti-tumour effect of interferon-α/5-fluorouracil in hepatocellular carcinoma cells. Br J Cancer 103:1617–1626
76. Zhou J, Zhou Y, Yin B, Hao W, Zhao L et al (2010) 5-Fluorouracil and oxaliplatin modify the expression profiles of microRNAs in human colon cancer cells *in vitro*. Oncol Rep 23:121–128
77. Shah MY, Pan X, Fix LN, Farwell MA, Zhang B (2011) 5-Fluorouracil drug alters the microRNA expression profiles in MCF-7 breast cancer cells. J Cell Physiol 226:1868–1878
78. Gumireddy K, Young D, Xiong X, Hogenesch J, Huang Q et al (2008) Small-molecule inhibitors of microrna miR-21 function. Angew Chem Int Ed Engl 47:7482–7484
79. Schmittgen TD, Jiang J, Liu Q, Yang L (2004) A high-throughput method to monitor the expression of microRNA precursors. Nucleic Acids Res 32:e43
80. Chang J, Nicolas E, Marks D, Sander C, Lerro A et al (2004) miR-122, a mammalian liver-specific microRNA, is processed from hcr mRNA and may downregulate the high affinity cationic amino acid transporter CAT-1. RNA Biol 1:106–113
81. Esau C, Davis S, Murray S, Yu X, Pandey S et al (2006) miR-122 regulation of lipid metabolism revealed by *in vivo* antisense targeting. Cell Metab 3:87–98

82. Parkin D, Bray F, Ferlay J, Pisani P (2005) Global cancer statistics, 2002. CA Cancer J Clin 55:74–108
83. Lin C, Gong H, Tseng H, Wang W, Wu J (2008) miR-122 targets an anti-apoptotic gene, Bcl-w, in human hepatocellular carcinoma cell lines. Biochem Biophys Res Commun 375:315–320
84. Gramantieri L, Ferracin M, Fornari F, Veronese A, Sabbioni S et al (2007) Cyclin G1 is a target of miR-122a, a microRNA frequently down-regulated in human hepatocellular carcinoma. Cancer Res 67:6092–6099
85. Fornari F, Gramantieri L, Ferracin M, Veronese A, Sabbioni S et al (2008) MiR-221 controls CDKN1C/p57 and CDKN1B/p27 expression in human hepatocellular carcinoma. Oncogene 27:5651–5661
86. Yang JD, Roberts LR (2010) Hepatocellular carcinoma: a global view. Nat Rev Gastroenterol Hepatol 7:448–458
87. Jopling C, Yi M, Lancaster A, Lemon S, Sarnow P (2005) Modulation of hepatitis C virus RNA abundance by a liver-specific MicroRNA. Science 309:1577–1581
88. Jopling C, Norman K, Sarnow P (2006) Positive and negative modulation of viral and cellular mRNAs by liver-specific microRNA miR-122. Cold Spring Harb Symp Quant Biol 71:369–376
89. Lanford R, Hildebrandt-Eriksen E, Petri A, Persson R, Lindow M et al (2010) Therapeutic silencing of microRNA-122 in primates with chronic hepatitis C virus infection. Science 327:198–201
90. Young D, Connelly C, Grohmann C, Deiters A (2010) Small molecule modifiers of microRNA miR-122 function for the treatment of hepatitis C virus infection and hepatocellular carcinoma. J Am Chem Soc 132:7976–7981
91. Yi M, Lemon S (2004) Adaptive mutations producing efficient replication of genotype 1a hepatitis C virus RNA in normal Huh7 cells. J Virol 78:7904–7915

Index

S. Alahari (ed.), *MicroRNA in Cancer*,
DOI 10.1007/978-94-007-4655-8, © Springer Science+Business Media Dordrecht 2013